RAJA RAMANNA
A Renaissance Man

RAJA RAMANNA
A Renaissance Man

Editors

Dinesh K Srivastava
V S Ramamurthy

National Institute of Advanced Studies, India

World Scientific

NEW JERSEY · LONDON · SINGAPORE · BEIJING · SHANGHAI · HONG KONG · TAIPEI · CHENNAI · TOKYO

Published by

World Scientific Publishing Co. Pte. Ltd.

5 Toh Tuck Link, Singapore 596224

USA office: 27 Warren Street, Suite 401-402, Hackensack, NJ 07601

UK office: 57 Shelton Street, Covent Garden, London WC2H 9HE

Library of Congress Cataloging-in-Publication Data

Names: Srivastava, D. K. (Dinesh Kumar) editor | Ramamurthy, V. S. editor
Title: Raja Ramanna : a renaissance man / editors, Dinesh K. Srivastava,
    National Institute of Advanced Studies, India,
    V.S. Ramamurthy, National Institute of Advanced Studies, India.
Description: Hackensack, New Jersey : World Scientific, [2026] |
    Includes bibliographical references and index.
Identifiers: LCCN 2025012319 | ISBN 9789819814428 hardcover |
    ISBN 9789819814442 ebook for individuals | ISBN 9789819814435 ebook for institutions
Subjects: LCSH: Ramanna, Raja, 1925-2004 | Nuclear engineers--India |
    Nuclear engineering--India | LCGFT: Festschriften | Festschriften
Classification: LCC TK9014.R35 R35 2026 | DDC 621.48092--dc23/eng/20250401
LC record available at https://lccn.loc.gov/2025012319

British Library Cataloguing-in-Publication Data
A catalogue record for this book is available from the British Library.

For any available supplementary material, please visit
https://www.worldscientific.com/worldscibooks/10.1142/14346#t=suppl

Desk Editors: Kannan Krishnan/Joseph Ang

Typeset by Stallion Press
Email: enquiries@stallionpress.com

To
The thousands of scientists, engineers and
technicians who strived to transform
the dreams of Dr. Raja Ramanna
into a vibrant reality

Foreword

Dr. Raja Ramanna, the founder director of the National Institute of Advanced Studies (NIAS), was a great visionary. During his five-decade-long service to the nation, he had built or transformed institutions, formulated policies for science and technology and above all showed the path for integrating science and technology with social science and humanities while keeping his passion for music alive. It is rare to come across such an individual who has contributed to the field of nuclear and defence research, as well as to philosophy, and is a legendary pianist. NIAS decided to celebrate his 100th Birth Centenary from January 2024 to 2025 by organising various activities at NIAS to pay our respects to the great inspiring and motivating leader.

The Celebrations were inaugurated by his close friend Prof. C. N. R. Rao, Honorary President, Jawaharlal Nehru Centre for Advanced Scientific Research (JNCASR), Bengaluru. It was followed by a Tribute Concert, "Paramanu and Paramatma" symbolising his contributions to nuclear research and philosophy through music, which was so dear to him.

He was the architect of India's Peaceful Nuclear Explosion in 1974 and brought laurels to the country. He was a firm believer that nuclear energy has to be a part of the electricity requirement of India. His vision has come true now, and the Government of India (GoI) has announced a major initiative for setting up small and modular reactors (SMRs) for the requirement of captive power for steel, aluminium and cement industries. NIAS has been actively involved in providing various inputs on SMRs as well as identifying possible sites for these.

To mark his vision about nuclear research, a Seminar on "Nuclear Physics Research in India" was organised and was inaugurated by Dr. Ajit Kumar Mohanty, Chairman of Atomic Energy Commission (AEC) and Secretary of Department of Atomic Energy (DAE), GoI. Many eminent scientists who were associated with him delivered talks during this Seminar. His contributions during his initial days at Tata Institute of Fundamental Research (TIFR) and later at Bhabha Atomic Research Centre (BARC) about nuclear fission and neutron scattering, setting up and nurturing of institutions, *viz*., Variable Energy Cyclotron Centre, Indira Gandhi Centre for Atomic Research, Atomic Energy Regulatory Board, Centre for Advanced Technology (later renamed as Raja Ramanna Centre for Advanced Technology) and a Training School for nuclear science and engineering for young scientists and engineers, and Indian Physics Association were remembered. Dr. Raja Ramanna was committed to using nuclear energy, especially for health care, and promoted the use of radioisotopes for diagnostics as well as for therapy. As a Chair of the Scientific Advisory Committee of the International Atomic Energy Agency (IAEA), he addressed concerns of the developing nations about nuclear energy and made outstanding contributions for planning various programmes and their implementation.

His contributions to defence research have been legendary. NIAS had organised a Seminar on "Emerging Technologies in Defence" to remember his vision about the Defence sector both as a Scientific Adviser to the Raksha Mantri and the Chief of the Defence Research and Development Organisation (DRDO) and later as a Raksha Rajya Mantri (Minster of State for Defence). Dr. V. K. Saraswat, Member of NITI Aayog, inaugurated the Seminar and recalled his association with Dr. Ramanna. Dr. Sameer Kamath, Chairman of Defence Research and Development Organisation (DRDO), Dr. S. P. Somanath, the then Chair of Indian Space Research Organisation (ISRO), and many other eminent scientists and technologists participated and articulated their vision about future defence requirements. Dr. Ramanna had reformed this organisation and initiated major programmes of weapon system development, Main Battle Tank, Integrated Guided Missile Development, Light Combat Aircraft (LCA) and many others. All these projects have now fructified and provided impetus to the development of the indigenous industry. The decision to design and build a nuclear-powered submarine, jointly executed by the

Indian Navy, DRDO laboratories and BARC, was fructified in 2016, and it has boosted the strategic capabilities of India. He created the Department of Defence Research and Development to provide scientific advice to the Ministry of Defence and armed forces. One of the major achievements was reforming the recruitment and assessment policies and the introduction of Flexible Complimenting Scheme for the promotion of scientists and engineers in DRDO.

As a parliamentarian and a minister, he showed his analytical skills and intellect for timely interventions and decision-making. He had a rare ability to communicate with the political leadership. He had played a vital role in rescinding the order of grounding of the entire fleet of Airbus 320 of Indian Airlines, after the tragic accident of Airbus 320 in Bengaluru.

He had a keen interest in philosophy and Yoga. As a part of Birth Centenary Celebrations, a panel discussion on "The Relevance of Traditional Wisdom in Modern Science" was organised. Eminent philosophers and scientists debated about the importance of traditional knowledge. Dr. Ramanna wrote several books, *viz. Mukundamala of Kulashekhara Alwar* (Translation), *Sanskrit and Science*, and *Years of Pilgrimage: An Autobiography* and important articles on philosophy, e.g., *Moksha: A Critique, Scientific Philosophy with reference to Buddhist Thought*.

In 1988, he was invited by Shri J. R. D. Tata to set up a research institute to promote multidisciplinary and interdisciplinary research to address societal issues by integrating science and technology with social science and humanities. He had shown an exceptional drive to adapt to the new challenge and NIAS owes his current eminence in this field due to the seeds sown by Dr. Ramanna. He initiated training programmes for the government and corporate sector to bring awareness about developments in science, technology, social science and arts to increase their ability for decision-making and thus contribute to the nation's progress. These training programmes are continuing even today and have brought immense visibility to NIAS. His legacy continues in NIAS through "Wednesday Discussions" and "Associates Meeting", as well as programmes on consciousness, security and strategic studies, education, social science, science and engineering and humanities.

The Foundation Day of NIAS was dedicated to Dr. Ramanna. The Foundation Day address was delivered by Dr. A. K. Mohanty, and he articulated the impact of Dr. Ramanna's contributions on the national scenario. Dr. Ramanna was deeply associated with the Karnataka Police Karnatak Band, and they performed as a part of this event. The Celebrations ended with the Dr. Raja Ramanna Memorial Lecture by Prof. Ajay Sood, Principal Scientific Adviser to the Government of India, who was one of the students at the BARC Training School. A video film on "Legacies of Innovation" specially made for this occasion, was also screened. A musical concert, "Stringing East-West Resonances: The Bangalore String Ensemble" was organised to pay tributes to this gifted pianist.

Music was the first love of Dr. Ramanna. He started learning piano at the age of six and mastered it by the age of twelve and performed in front of the Maharaja of Mysore. Dr. Homi Bhabha, a great musician himself, saw Dr. Ramanna playing piano at one of the concerts in Mysore. This incident changed the career path of Dr. Ramanna and he became a nuclear scientist. But his interest in music continued till the end and he gave many concerts and mentored many youngsters. He was instrumental in setting up the Bangalore School of Music. His book *The Structure of Music in Ragas and Western Music* is a scholarly contribution to music and is highly acclaimed.

Dr. Ramanna was a multifaceted personality: a brilliant scientist, an accomplished technologist, an inspiring leader, an institution builder, a philosopher of repute, an able administrator, a connoisseur of arts and above all a doyen of Western music. He was simple and very modest in his personal life and professional career.

Dr. Dinesh Srivastava and Dr. V. S. Ramamurthy have painstakingly collected, documented and edited this book, *Raja Ramanna: A Renaissance Man*, with reminiscences of people who have been closely associated with Dr. Ramanna. I am extremely grateful to them for their effort and time. I am sure this book will serve as a record of the achievements of Dr. Ramanna and inspire many young scholars to dedicate their services to the country and humanity.

Shailesh Nayak
Director
National Institute of Advanced Studies (NIAS)
Bengaluru

About the Editors

Dinesh Kumar Srivastava (born in 1952) is presently an Indian National Science Academy Senior Scientist and Honorary Visiting Professor at the National Institute of Advanced Studies, Bengaluru. He is now working on energy, environment, climate change, international collaborations, and science outreach along with a continuation of his research on quark–gluon plasma.

He obtained his graduation from Allahabad University in 1970 and joined the Training School of the Bhabha Atomic Research Centre, Mumbai. He started working at the Variable Energy Project of Bhabha Atomic Research Centre in 1971 and retired as the Director and Distinguished Scientist at the Variable Energy Cyclotron Centre, Kolkata, in 2016. Later, he continued there as a Department of Atomic Energy (DAE) Raja Ramanna Fellow until 2019. He worked at the National Institute of Advanced Studies, Bengaluru, as a Homi Bhabha Chair Professor, during 2019–2023. He is a Fellow of the National Academy of Sciences, India, and the Indian National Science Academy.

His book *Climate Change and Energy Options for a Sustainable Future* written with V. S. Ramamurthy was published by World Scientific Publishing Company (WSPC) in early 2021. Another one *Art and Science of Managing Public Risks* by V. S. Ramamurthy, Dinesh Kumar Srivastava and Shailesh Nayak was also published by WSPC in early 2022. A new book exploring the importance of international collaborations, *Science Beyond Borders: International Collaborations in Basic Sciences*, written by Dinesh Kumar Srivastava, V. S. Ramamurthy. Shailesh Nayak and D. Suba Chandran was published by WSPC in early 2025.

V. S. Ramamurthy is a well-known Indian nuclear scientist with a broad range of contributions from basic research to science administration. Prof. Ramamurthy started his career at Bhabha Atomic Research Centre, Mumbai, in the year 1963. He has made important research contributions, both experimental and theoretical, in many areas of nuclear fission and heavy ion reaction mechanisms, statistical and thermodynamic properties of nuclei, physics of atomic and molecular clusters, and low-energy accelerator applications. During the period 1995–2006, Prof. Ramamurthy was fully involved in science promotion in India as Secretary to the Government of India, Department of Science and Technology (DST), New Delhi.

He was also the Chairman of the IAEA Standing Advisory Group on Nuclear Applications for nearly a decade. After retirement from government service, Prof. Ramamurthy, in addition to continuing research in Nuclear Physics in the Inter-University Accelerator Centre, New Delhi, has also been actively involved in human resource development in all aspects of nuclear research and applications. Prof Ramamurthy is also a Chairman of Recruitment and Assessment Board, Council of Scientific and Industrial Research, and a Member of National Security Advisory Board. In recognition of his services to the growth of Science and Technology in the country, Prof. Ramamurthy was awarded one of the top civilian awards of the country, the Padma Bhushan, by the Government of India in 2005. Prof. Ramamurthy was the Director of the National Institute of Advanced Studies from 2009 to 2014 and is now an Emeritus Professor there.

His book *Climate Change and Energy Options for a Sustainable Future* by Dinesh Kumar Srivastava and V. S. Ramamurthy was published by World Scientific in early 2021. Another one *Art and Science of Managing Public Risks* by V. S. Ramamurthy, Dinesh Kumar Srivastava and Shailesh Nayak was also published by World Scientific in early 2022. A new book exploring the importance of international collaborations, *Science Beyond Borders: International Collaborations in Basic Sciences*, written by Dinesh Kumar Srivastava, V. S. Ramamurthy. Shailesh Nayak and D. Suba Chandran, was published by WSPC in early 2025.

Contents

Chapter 1

Dr. Raja Ramanna: A Sacred Memory

R. Chidambaram

DAE Institutes of Higher Learning, Mumbai, India,

Scientific Adviser to Government of India, Mumbai, India,

Department of Atomic Energy, Mumbai, India,

Atomic Energy Commission, Mumbai, India

डॉ. आर. चिदंबरम
Dr. R. Chidambaram

भारत सरकार
Government of India

अध्यक्ष (मानद)
एजवि प्रगत अध्ययन ईस्थान
Chairman (Honorary)
DAE Institute of Advanced Studies
भारत सरकार के पूर्व प्रधान वैज्ञानिक सलाहकार,
पूर्व अध्यक्ष, परमाणु कार्या आयोग
Former Principal Scientific Adviser to GOI,
Former Chairman, Atomic Energy Commission

Dr. Raja Ramanna was hand-picked by Dr. Homi Bhabha to start the physics programme in TIFR (along with Dr. M G K Menon) and later the physics programme in BARC. He was a very important person in my career. I remember when I joined the neutron diffraction programme in CIRUS, the expert from TIFR asked me why I was changing from NMR (in which field I had my Ph.D. from the Indian Institute of Science) to Neutron Diffraction. I responded that I was not changing my field – I was still very much interested in hydrogen bonding in solids; I was only changing the technique of studying the same problem because neutron diffraction gave direct information of hydrogen positions. While this debate was going on with the expert, Dr. Ramanna sat there smiling; of course, I got selected to BARC.

The other career – altering step in my career was also because of Dr. Ramanna. One day in 1969 he called to my office and asked me to work on shock waves and nuclear weapon design. Dr Nag Chauduri, who was heading the DRDO programme, was a good friend of Dr. Raja Ramanna, because both were nuclear phys cists. That is how we could carry out the 1974 nuclear test - the so - ca led Peaceful Nuclear Explosion Test. The tests in 1998 were, of course, nuclear weapon tests, and India declared itself as a Nuclear Weapon State. We must appreciate the far - sightedness of Dr. Ramanna.

To conclude, Dr. Raja Ramanna was one of the nuclear pioneers and founded the physics part of the nuclear programme. We must be grateful to him for helping make India a strong nuclear state, both on the development and the security side.

R. Chidambaram

(R Chidambaram)

Dr. Rajagopal Chidambaram (11 November 1936–4 January 2025) was a doyen of science and technology whose contributions furthered India's nuclear prowess and strategic self-reliance. He had his early education in Meerut and Chennai and completed his B.Sc. with honours in physics. As he stood first at the departmental and the university level of Madras University in 1956, he was invited to teach introductory physics laboratory courses while studying for M.Sc. He wrote a fundamental thesis on analogue computers in 1958 for his M.Sc., joined the doctoral programme of the Indian Institute of Science and was awarded the Ph.D. in 1962. His thesis contained research work on the development of nuclear magnetic resonance and was conferred with the Martin Forster Medal for the best doctoral thesis submitted to the Indian Institute of Science.

He joined the Bhabha Atomic Research Centre in 1962 and started working on crystallography and condensed matter physics, using neutron scattering – developing all the equipment. In 1967, he started his work on the physics, design and metallurgical aspects of plutonium under high pressure on suggestions from Dr. Raja Ramanna.

Dr. Chidambaram played a pivotal role in shaping India's nuclear capabilities and an integral role in the nation's first nuclear test (Smiling Buddha, Pokhran-I) in 1974 and led the Department of Atomic Energy team during the nuclear tests (Operation Shakti, Pokhran-II) in 1998. His contributions established India as a nuclear power on the global stage.

Dr. Chidambaram's research in high-pressure physics, crystallography and materials science significantly advanced these fields. After the test of the nuclear device at Pokhran-I in 1974, he started "open research" in high-pressure physics, and a complete range of instrumentation such as diamond anvil cells and gas guns for launching projectiles were indigenously built. He also laid the foundation of theoretical high-pressure research for the calculation of the equation of state and phase stability of materials by first-principles techniques. His pioneering work in these areas laid the foundation for modern materials science research in India. He and his colleagues worked out the equation of the state of plutonium, which is still classified by all nuclear weapon states. When technology denials and sanctions were imposed on India, he spearheaded the programme to develop supercomputers.

Dr. Chidambaram was a visionary leader who believed in utilising the power of science and technology to drive national development. He championed initiatives in areas such as energy, healthcare and strategic self-reliance, and he steered numerous projects that significantly advanced India's science and technology landscape. He was instrumental in initiating India's indigenous development of supercomputers and conceptualising the National Knowledge Network (NKN), which connected research and educational institutions across the country.

He was an ardent advocate of applying science and technology to national development and established programmes such as the Rural Technology Action Groups (RUTAG), Society for Electronic Transactions and Security (SETS) and Core Advisory Group for R&D in the Automotive Sector (CAR) to increase academia–industry interaction and emphasised "Coherent Synergy" in India's scientific efforts. Dr. Chidambaram also emphasised the importance of "Directed Basic Research" as an addition to (and not a substitute for) self-directed basic research.

Dr. Chidambaram held numerous prestigious roles throughout his illustrious career, including Principal Scientific Adviser to the Government of India (2001–2018), Director of the Bhabha Atomic Research Centre (1990–1993) and Chairman of the Atomic Energy Commission and Secretary to the Government of India, Department of Atomic Energy (1993–2000).

He was the Chairman of the Board of Governors of the International Atomic Energy Agency (1994–1995). He also served as a member of the IAEA's Commission of Eminent Persons, contributing to the organisation's vision for 2020 and beyond.

He was the President of Shree Chitra Tirunal Institute of Medical Sciences Thiruvananthapuram, the Chairman of the Board of Governors of the Indian Institute of Technology Bombay (1994–1997) and Indian Institute of Technology Madras (2008–2011), the Chancellor of the University of Hyderabad and the Chairman of the Board of Governors of Indian Institute of Technology Jodhpur. Dr. Chidambaram was a Member of Space Commission (2009–2014). During 1990–1999, he was a member of the Executive Committee of the International Union of Crystallography and the last three years as its Vice President.

Dr. Chidambaram was a recipient of many Awards including Padma Shri (1975) and Padma Vibhushan (1999) and was conferred Honoris Causa DSc by several universities. He was a Fellow of all the major

Science Academies in India including the National Academy of Engineering and "The World Academy of Sciences" Trieste.

[Dr. R. Chidambaram enthusiastically supported our plan to bring out this monograph of memoirs of close colleagues of Dr. Raja Ramanna and readily sent us his contribution on 11 January 2025. We deeply mourn his passing away during the preparation of this. (Eds.)]

Chapter 2

Dr. Raja Ramanna: My Mentor

Anil Kakodkar

Homi Bhabha National Institute, Mumbai, India

Rajiv Gandhi Science & Technology Commission, Mumbai, India

Atomic Energy Commission, Mumbai, India

kakodkaranil@gmail.com

I have been very fortunate to come close to Dr. Ramanna and benefit from his guidance, mentoring and wisdom. When I joined Bhabha Atomic Research Centre (BARC, then Atomic Energy Establishment, Trombay) in August 1964, after a one-year orientation programme at the 7th batch of BARC Training School, I had earned a fair bit of visibility, having passed out of the Training School with a record performance of having scored the highest marks among all batches and all disciplines till then. The record remained unbroken for another seven years. The Training School was under the overall leadership of Dr. Raja Ramanna who was among the top five directors of BARC led by Dr. Homi Jehangir Bhabha. We were told that Training School was among the first critically important activities that Dr. Bhabha had initiated keeping the human resource needs of the emerging atomic energy programme.

The Training School has contributed to nearly the entire talent that has gone into the making of our atomic energy programme. Talent nurture was presumably a matter of considerable importance and priority for the top leadership then. A good part of it was also involved in teaching and academic management of the Training School activities. I also had my share of these activities in earlier years. Two important assignments that are worth mentioning were preparing a Training School-like programme for Egypt which had sought India's help and defining human resource needs and related action plan for India's nuclear power programme that was prepared at the instance of Dr. Vikram Sarabhai who had succeeded Dr. Bhabha as Chairman Atomic Energy Commission (AEC) after his sad demise. Both these activities were under the overall guidance of Dr. Ramanna, and I was told that they were well appreciated.

While I was at Nottingham University for my higher studies, I was informed that a plan to build a pulsed fast reactor was shaping up and that I would have to develop a large rotor that would sweep a reflector block past the reactor core at high speed to create a reactivity pulse making the reactor momentarily prompt supercritical. I was asked to start preparing to take up this assignment immediately on my return. The project was led by Dr. P. K. Iyengar and was a part of the activities of the Physics Group led by Dr. Ramanna. I did start working on this project on my return along with other activities related to developing Pressurised Heavy Water Reactor (PHWR) components and systems. The project did not go very far, and I remained busy with the PHWR programme which really was the core of my work.

Sometime around 1973 I was assigned to develop DHRUVA reactor in addition to my other ongoing activities and was working very closely with our director Shri V. N. Meckoni. One day he asked me to see Dr. Ramanna who was then Director of BARC. I telephoned Director BARC's office and sought an appointment. The next day Shri Meckoni again enquired why I had not seen Dr. Ramanna. To my reply that I had already sought the appointment, he suggested that I should just go and drop in. I did that. Dr. Ramanna was visibly angry at my being so protocol-conscious and told me to be informal and see him frequently including visiting him at home. He even said that I did not have to have a reason to see him. This was the time he wanted to induct me into the Peaceful Nuclear Explosion (PNE) experiment programme. He even asked me if I had the habit of talking while in sleep.

One day while at Pokhran, there was some complication while assembling the PNE device. That caused a lot of delay in returning to our base mess. It was late, nearly midnight. When we reached, I found Dr. Ramanna anxiously waiting for us. He enquired what was wrong. It was difficult to explain the nitty-gritty of a complex mechanical assembly to a non-mechanical engineer. So, I simply assured him *ho jayega* implying that the problem would be solved and he need not worry. He appeared amused, left the discussion there and addressed me as "*ho jayega* man" for several months thereafter. Dr. Ramanna never believed in micro-management and rather moved forward based on trust in his colleagues. His judgement about people rarely went wrong.

A little after 1974 PNE, Dr. Ramanna moved to the Defence Research and Development Organisation (DRDO). I had a few occasions to meet him at some public functions. I used to be quite embarrassed when he would make me sit with him although his hosts would want his full attention. But that was Dr. Ramanna who was least concerned about unnecessary protocols. He returned to BARC after a few years, this time as both Director of BARC and Secretary, Department of Atomic Energy (DAE), Government of India. DHRUVA project had advanced quite a bit by then. There was a case of cost escalation in one of the contracts. This required approval of the Internal Finance Adviser (IFA) followed by his approval as Director BARC and Secretary, DAE. I had made the necessary paper and kept it with me to get IFA's approval first whenever I saw him. While casually briefing Dr. Ramanna on DHRUVA's progress, I did mention this case and mentioned that the paper would come for his approval after the IFA's signature. Dr. Ramanna sensed that the paper was with me. He asked me to show it to him, read it and despite my telling him that he should await IFA's signature, signed it off. To pacify me, he explained at length how one must have full faith in one's trusted colleagues if one wants to achieve big things. He also asked me to convey to IFA that even though he as Secretary had approved the note, he was at liberty to make his observations. During the construction phase of DHRUVA, there were several technical controversies that needed resolution. I learned a lot about scientific leadership from Dr. Ramanna at that time. Every time he asked me to confidently go ahead and defended his colleagues on either side of the divide without himself taking sides.

One unique innovation that Dr. Ramanna implemented during his directorship at BARC was the setting up of the Informal Director's Advisory Committee (IDAC). The group consisted of twenty-odd high-performing young scientists of BARC who were not holding any formal position in BARC's administrative setup. The committee would take up issues they considered important and come up with recommendations for him. He himself used to actively participate in the discussions. This had a two-way benefit. We all were getting groomed in science management while he would get useful feedback. I benefitted a lot from these discussions.

Dr. Ramanna's concern about progress in my career deeply impacted me. Once I was asked by his office to rush to the Central Complex where the selection committee for senior promotions was in session. After some time, as I was waiting in the adjacent lounge, Dr. Ramanna came up to me, apologised and asked me to return. Obviously, something he tried, did not work out. For me, that Dr. Ramanna cared so much about me was a matter of greater happiness. Sometime later, I was again called to appear before the selection committee. I was waiting along with other candidates in the room adjoining the selection committee room. Dr. Ramanna came in, called me out and took me to his room nearby only to tell me to remain calm and confident. This happened twice on the same day. Obviously, there were differences in the selection process that Dr. Ramanna was tense about and wanted to ensure that I did not become a casualty. My interview was over in five minutes with a positive outcome. But again, Dr. Ramanna's concern for me was overwhelming.

Dr. Ramanna became Chairman of AEC in 1983. India's nuclear power programme was primarily based on 220 MWe PHWR units till then. He initiated two important things: One was to define a long-term nuclear power programme for the country and the second was to take up the design and development of 500 MWe PHWR unit. I was closely involved in both. The committee to develop the nuclear power programme was set up under the chairmanship of Dr. M. V. Ramaniah in which I was a member. After a very detailed assessment, the committee proposed a programme of building 10,000 MWe nuclear capacity by the year 2000. This was a challenging ask but essential, considering the long-term energy needs of the country. I wish everyone who followed Dr. Ramanna was as bold as him. It had become clear while working for the committee that the key challenge before the country was the lack of adequate uranium resources and that our progress on the thorium front was far too slow.

On 500 MWe PHWR development, Dr. Ramanna was keen on greater synergy between BARC and the Nuclear Power Board. He indeed instructed me about this, now that DHRUVA was about to be completed. He even created space for the 500 MWe group within BARC. It seemed to me that the issue finally hinged on to whom I would report. This was the least of the issues for me since I had worked under multiple bosses simultaneously more or less throughout my career. Dr. Ramanna was a liberal and valued borderless working. The issue, however, could not be resolved between BARC and the Nuclear Power Board. Mine and my group's involvement in the development of the 500 MWe PHWR system remained unaffected and continued the same way as it was in the case of 220 MWe units. In the process, we both came even closer.

One day Shri Shivshankar Menon who was in DAE at that time told me that Dr. Ramanna wanted me to join him on his Cairo–Vienna–Warsaw tour that was being scheduled. He gave me inputs on ongoing activities with Egypt, International Atomic Energy Agency (IAEA) and Poland as well as other relevant details from his database. That however did not tell me what was expected of me. Nor could Menon help me with it. There were other senior scientists who were involved in the ongoing cooperation and were in any way a part of the delegation to accompany him. I then asked Dr. Ramanna himself about what was expected of me so that I could prepare for the tour properly. His reply was, "I am asking you to come with me, just come".

We were sitting down for breakfast on our first day in Cairo. Dr. Ramanna was not very keen on the usual stereotype menu. He was looking for some green chillies. Our host had chillies in his kitchen garden. He rushed to his house nearby and brought a bagful of chillies. Dr. Ramanna was happy. He made a few sandwiches with chillies and enjoyed his breakfast. He asked me to keep the bag of chillies with me. Our next halt was Vienna. Ambassador Jagadish Hiremath hosted a reception for the diplomatic core in Vienna in honour of Dr. Ramanna. The suggestion of Dr. Ramanna to serve chilli pakoras at the reception was a great hit.

At Vienna, Dr. Ramanna was to attend the Scientific Advisory Committee to the IAEA Director General. He got a chair placed for me next to him. Listening to the proceedings of the meeting itself was a great learning experience. Further, Dr. Ramanna's interventions were quite out of the box. He was clearly the centre of attraction for the whole meeting. He had something more to tell me after every intervention about the underlying currents and how we should be bold in protecting our interests. It was really a hands-on tutorial for me. I realised then that a Regional Cooperation Agreement among Asian countries was actively nurtured by India and was perhaps the most successful regional nuclear cooperation programme in which India had earned a lot of respect.

At Warsaw, other senior scientists from India joined us for the bilateral meeting. After the first introductory meeting, there was an extensive tour of various laboratories. Dr. Ramanna had seen these labs earlier, so he decided to rest and asked me to go around the labs with the group. When we sat down for the final meeting, Dr. Ramanna suddenly felt the need to summarise what the Indian delegation saw and where there is scope and interest in bilateral cooperation. I was sitting next to him. He placed a writing pad between the two of us and asked me to start jotting down points relating to the lab visits. He made an excellent uninterrupted speech looking at the points I was scribbling on the writing pad. He noticed my very disturbed and fearful state while all this was happening and patted me on the back at the end for the good job done. That incident taught me to remain calm and fearless and strengthened my confidence in being so.

At Cairo, there was some time after official engagements. Dr. Ramanna decided to visit the Tutankhamun Museum. One of his students who was working in Egyptian Atomic Energy volunteered to accompany us and wanted to arrange a guide to show us around. Dr. Ramanna was well familiar with the museum, had visited it earlier and had seen it up to a point. Not only that he did not need a guide, but he also took both of us to the point up to which he had seen in his last visit, and we moved further with Dr. Ramanna talking about the exhibits in full detail. As we finished, he even joked that if not anything, he could at least become a guide at Tutankhamun. On the way back from Tutankhamun, he emphasised how important it was to learn history and how such museums are a treasure house for learners.

As Dr. Ramanna was nearing superannuation, discussions on shaping National Institute of Advanced Studies (NIAS), Bengaluru, were taking place. I used to go to NIAS quite often during the days of Dr. Ramanna. In one such discussion at NIAS on how science programmes should be shaped, after a long brainstorming within a large and broad-based group, Dr. Ramanna simply poured cold water on the final conclusions of the group. I did not quite understand as I found the group conclusions to be quite logical. After the meeting, I went up to him and asked him the reason. He simply said that things in real life are complex and do not comply with seemingly straightforward logic. On some other occasion, I asked his advice on whether to move forward on a research collaboration proposal by a foreign laboratory. He enquired, "How strong are we in the area?" When I said that we were quite strong, he simply asked me to go ahead. He said, "if we are very weak, we are likely to be exploited. On the other hand, if we are strong, the collaboration will make us stronger". There was something deeply heuristic about Dr. Ramanna. At times, it was difficult to fathom the way he used to think. But his advice always had a lot of value.

One of the areas of NIAS's engagement concerns track 2 dialogues. While this was an important part of the NIAS portfolio, Dr. Ramanna had been extremely careful and at times circumspect. I have personally witnessed his predicaments in running such programmes and have learnt a lot from him in this important area.

When I became Chairman of the AEC, he invited me for dinner at the Taj Hotel, in Mumbai. Quite unlike other occasions, this time, the mentoring was about how I should conduct myself in Delhi and steer through the complex narratives that prevail. One day, he suddenly decided to drop in at my house in Mumbai. He seemed very concerned about me because I was pushing the high-stakes programme very fast. He advised me to be watchful about happenings around me. While I assured him about the sound approach behind anything that I was doing and being watchful about things around me, I also told him how the time was running out and why speedy actions were needed. Nevertheless, I was overwhelmed by his concerns about my well-being.

I had a similar experience earlier when I was working on DHRUVA. He seemed concerned by my nature as a workaholic. He often would suggest my devoting time to other aspects of life, such as sports, arts and culture. On the other hand, while working on DHRUVA, I felt that I should go through some management courses to be in a better position to manage large programmes. This was quite a routine matter in those days. I had even secured an admission. Dr. Ramanna was however not very happy about it. We talked about it at length. He was not very enamoured with the so-called American management ideas and felt that there are better things in life on which to devote time. I on my part did stress on a more organised way of working that management training may bring about. The discussion was not going anywhere. Finally, deferring to his wishes, I withdrew from the management program but with a condition that as a matter of principle he would in fact make it easier for other BARC officers to enrol for such programmes. He did keep his word and got a Trombay Council decision done on the issue.

Dr. Ramanna's love for music has been well known. I have heard his formal piano concerts. I could not do much about his advice to me about the importance of arts and culture in our lives. However, I witnessed his happiness when he was in the company of musicians while in Vienna attending IAEA meetings. One such family was Prof. Klaus-Peter Satler and Viloo Satler. Viloo, being of Indian origin, made it a point to meet me when I was in Vienna and often gave a set of music books for

Dr. Ramanna. Carrying these books to Dr. Ramanna had become a very enjoyable routine for me. When Dr. Ramanna left for his heavenly abode, I was in Vienna holding music books given by Viloo for Dr. Ramanna. I could reach them to his family but taking them further was beyond my reach.

My respects to my mentor.

Dr. Anil Kakodkar, FNASc, FNA, FNAE, joined the Bhabha Atomic Research Centre (BARC) in 1964 and became Director of BARC in the year 1996 and was Chairman, Atomic Energy Commission and Secretary to the Government of India, Department of Atomic Energy, during the years 2000–2009. He was DAE Homi Bhabha Chair Professor from January 2010 to January 2015 and INAE Satish Dhawan Chair of Engineering Eminence from January 2015 to January 2017. At present, he is Chancellor of Homi Bhabha National Institute, Chairman of the Rajiv Gandhi Science & Technology Commission and member of Atomic Energy Commission. He has devoted his life to the development of the atomic energy programme in India.

He was involved in the first successful Peaceful Nuclear Explosion Experiment conducted by India in 1974 at Pokhran and played a key role in the series of successful Nuclear Tests conducted in 1998. He has been awarded the Padma Shri, Padma Bhushan and Padma Vibhushan.

Chapter 3

Dr. Raja Ramanna: As I knew him Some Glimpses

G. Venkataraman

Sri Sathya Sai Institute of Higher Learning

Prashanti Nilayam - 515 134, Vidyagiri

Sri Sathya Sai District, Andhra Pradesh, India

gvraman108@gmail.com

In June 1955, I responded to an advertisement from the Tata Institute of Fundamental Research (TIFR), inviting applications for the post of Research Assistant. I was asked to appear for an interview on 4 July. The interview was conducted in a room in what used to be the Yacht Club during British days. Subsequently, the Yacht Club moved to a new location, and the building where my interview took place was taken over by the Department of Atomic Energy (DAE) and came to be known as the Old Yacht Club (OYC). This was the first time I had applied for a job and faced an interview committee. Seated at the table were a large number of men. I had never seen any of them except the Chairman, Prof. Bernard Peters.

Years earlier when I was studying in the Madras Christian College, Prof. Peters then at the University of Rochester in America came to our college campus to conduct balloon flight experiments related to the study of cosmic rays; this study was being done in collaboration with TIFR. Madras Christian College was far from the city and there was a large cricket ground on the campus which was suited for balloon launches. One of the members accompanying Prof. Peters was the professor of Physics at Wilson College in Bombay. Wilson College was a sister college of my college, and I believe that is how my college became the launch pad for Prof. Peters' balloon flight experiments.

Peters needed student volunteers for his experiments and since I was interested in physics, I readily offered to be one. Shortly before Peters and his group left, there was a party for the volunteers, and I took the opportunity to talk to him. At that time, there was a strong anti-American campaign against people who were considered to be communists. Many journalists, actors and university professors were the targets of this witch hunt and Prof. Peters happened to be one of them. Peters told me that when he was a student in Berlin, he had participated in riots against Hitler who was just coming to power in Germany. Like many others, young Peters fled to America and studied under Oppenheimer. Since Oppenheimer himself was under a cloud, Peters automatically became a target. Luckily for him, Homi Bhabha, the man who propelled India to the Atomic Age, offered Peters a position as a professor in TIFR. That is how Prof. Peters came to be the Chairman of the Committee which interviewed me.

My father was in Bombay at that time, and one of his colleagues knew a member of the Interview Committee. He told me, "Come with me. I know A. S. Rao, one of the professors in TIFR. He must have been in the Committee. He would be able to tell whether you have been selected or not". When we met Prof. Rao, he said, "You have been selected, and you will work with Dr. Ramanna. I suggest you meet him".

I then met Dr. Ramanna and came to learn that he too studied physics at Madras Christian College during the war years. He was also a student leader there. Dr. Ramanna had many friends in that college, and he regularly visited it. Once when I was working in the Reactor Research Centre (RRC) in Kalpakkam, I went to Madras airport to receive him. This was in the eighties. When I met him, he said, "Let us first go to our college; after all it is on the way". What I am trying to convey is that Ramanna maintained his contacts with the college and the older members of staff, who were his personal friends.

Dr. Ramanna was a friendly person and moved easily with people at all levels. He had a soft corner for me because I was not merely an alumnus of Madras Christian College but also on the same wavelength with him on many matters. That being the case, there is a lot I can say presently. However, due to limitations of age, I shall restrict myself to five contributions of Dr. Ramanna, which are not well known. More specifically, I shall be dealing with the following topics:

1. the role played by Dr. Ramanna in establishing the Indian Physics Association,
2. the initiative Dr. Ramanna took in ensuring that the Saha Institute for Nuclear Physics in Calcutta was fully funded by the Department of Atomic Energy,
3. the role played by Dr. Ramanna in giving shape to the Thrust Area Program in Physics, sponsored by the Department of Science and Technology (DST). Prof. M. G. K. Menon, once of TIFR, was Secretary of DST at that time,
4. the special effort made by Dr. Ramanna to persuade the DAE to take under its umbrella, several small institutions engaged in research in mathematics and theoretical physics,
5. towering all of the above was the key role he played in ensuring that India's Nuclear Power Programme was based on natural uranium-fuelled, heavy water moderated reactors rather than on light water reactors that used enriched uranium as fuel. This vital contribution of Ramanna is little known, and I heard about it directly from Dr. Ramanna himself. This contribution is so important I feel it **has** to be placed on record.

Let me now amplify the points made above, in the order listed.

Raja Ramanna and the Indian Physics Association (IPA)

Nobel Prize winner physicist Rabi of Columbia University in America once said (paraphrase):

Mathematics is like an imperious Queen who is seated on a throne. To the left of the Queen are her knights, the first being Physics. All other sciences are to the left of Physics!

This clearly is an egoistic declaration, but there is some truth to it.

When India was ruled by the East India Company, science as we now know it did not exist in the country. However, thanks to Lord Macaulay, public schools were started all over the country and English was the medium of instruction in many of them. For Macaulay, it was cheaper to teach English to Indians and hire them as clerks, than paying much higher salaries to Englishmen to do the same job.

In 1858, the British Government in London took over the governance of India from the East India Company. Queen Victoria ruled Britain at that time, and that is how India became the crown jewel of the British Empire. The British rulers of India promptly established three universities. One was in Calcutta (now Kolkata) which was then the capital of India, while the others were in Madras (now Chennai) and Bombay (now Mumbai). Following the English tradition, the British Government in India also set up three "feeder" colleges in Calcutta, Madras and Bombay. While the colleges in Calcutta and Madras were referred to as Presidency College, that in Bombay was referred to as Elphinstone College, Elphinstone being the name of the Governor of Bombay Presidency at that time.

It is thanks to those schools, colleges and universities that India produced Srinivasan Ramanujan, C. V. Raman, J. C. Bose, S. N. Bose, Meghnad Saha and Homi Bhabha, the man who took India to the atomic age well before many countries in Europe. What I am trying to say is that modern science and mathematics came to India only after 1857, when the British began to rule India.

Move forward now to 1960 or so. At that time, the physics community in India was small and most of us knew each other. Where scientific organizations in India are concerned, the oldest among them is the Indian Science Congress. Subsequently, three academies of science came into existence, namely, the Indian Academy of Science (IAS), the Indian National Science Academy (INSA) and the National Academy of Sciences India (NASI). They all have their own histories, but I shall not go there. What I would like to say is that the physics community did not have its own distinctive scientific organization such as, for example, the American Physical Society (APS).

Recognising this, one year when the annual DAE Science Symposium met in Roorkee, Dr. Ramanna organized a meeting where it was resolved to form and register the Indian Physics Association (IPA). I promptly became a life member. Once the IPA came into existence, it began to publish a quarterly journal titled *Physics News*. One may say *Physics News* was the Indian version of *Physics Today* published by APS. The difference was that unlike the APS, which is wealthy, the IPA was born poor and stayed poor for a long time. No wonder that finding money to publish the quarterly journal became a struggle right from the beginning.

I was the President of IPA in 1988, which was celebrated as the Raman Birth Centenary Year. I was desperate to bring out a good edition of *Physics News* but had no cash. I had to struggle very hard to raise the required money and do justice to both Raman as well as Raman Effect. I am happy to say that currently *Physics News* is doing well and that the articles in it, all written by the young and active physicists of today, are quite impressive. I am particularly happy to note that there are now plenty of women physicists. Equally important is the fact that currently the IPA has established contacts with physics bodies all over the world, particularly those in Japan, South Korea, China, Singapore, Indonesia and Australia. And, thanks to Zoom, IPA now regularly arranges lectures by eminent physicists from all over the world.

Raja Ramanna Extends a Helping Hand to Saha Institute

I shall take it for granted that readers know something about Meghnad Saha. He was a product of Calcutta University and slightly younger than Raman. His claim to fame came from what is known as the Saha Ionization Formula, details about which can easily be found by googling. It suffices to say that Saha's work provided an important key to astronomers, at a time when they were trying hard to characterize and catalogue stars.

Besides physics, Saha was interested in many things, ranging from rivers to Mesopotamian culture and politics. After Independence, he contested as an independent and got elected as a Member of Parliament. Although Saha and Nehru (India's first Prime Minister) were political adversaries, Nehru asked Saha to deliver lectures to members of Parliament on science. Both Saha and Nehru strongly believed that India could progress only through science and technology.

During the period before World War II (when India was ruled by the British), Saha met prominent astronomers in America, all of whom knew about his important contribution. However, no one was interested in hiring him as a visiting professor, which is unfortunate. Following the discovery of the neutron in 1932, nuclear physics became a hot topic, and Saha promptly became interested in that subject. He had seen the cyclotron built by Lawrence in Berkeley and tried to build one in Calcutta with help from Lawrence. Meanwhile, World War II broke out and Saha's attempt to construct a cyclotron came to a halt.

When the war commenced, Europe automatically became the theatre of war (on the Western front) and science in Europe came to an abrupt halt. Where America was concerned, physicists seemed to have suddenly vanished. It was only after the end of World War II that the world knew that the atom bomb and radar projects had sucked all leading physicists in America into war work. World War II ended in August 1945, and slowly scientists in America began to get back to universities. Understandably, the situation was not that simple in Europe since the bulk of the war on the Western Front was fought in Europe. Simply put, a large part of Europe was in ruin.

Winston Churchill was the wartime Prime Minister of Britain. However, such was the nature of British politics that Churchill was defeated in the elections held in July 1945, and Clement Atlee of the Labour Party became the Prime Minister. Atlee realized that Britain was financially broke and the very first thing that he did was to make India independent. For Atlee, maintaining India as a crown jewel of the British Empire was too expensive! India became independent on 15 August 1947, with Jawaharlal Nehru as its first Prime Minister. I was studying in school then and I still recall that day!

Nehru appreciated not only the huge role played by science and technology during the war but also the importance of atomic energy. He wasted no time and formed an Atomic Energy Commission (AEC) as early as August 1948. Bhabha quickly got into the act, and since he could dream big, he promptly got Nehru excited about atomic energy. The latter gave Homi Bhabha a free hand and history shows that Bhabha delivered the goods.

I mention all this because Saha bitterly opposed what Bhabha was doing. Saha's view was that India should enter the atomic age via research in universities. Saha could not be blamed because "big science" did not exist before World War II, and he had little idea of what it meant to build a nuclear reactor. Britain had that know-how, and those who were engaged in that work were all Bhabha's Cambridge buddies! In 1947, the Canadians built their first nuclear reactor, and guess who the mastermind of all that was! Cambridge physicist, W. B. Lewis! I am going into all this because it has a bearing on the gesture Dr. Ramanna made later to the Saha Institute for Nuclear Physics (SINP).

Ramanna was a nuclear physicist, and it is no surprise that he wanted India to have a good cyclotron. There was no space in Bombay to build a cyclotron-based research centre. However, in Calcutta, there was a huge area known as Salt Lake close to the airport where plenty of land was available and at low cost. Ramanna felt this would be an ideal place to locate an accelerator. He wanted nuclear physicists from all over the country to do experiments at the cyclotron. Besides, Calcutta had produced many good scientists in earlier times. Above all, Ramanna wanted minimal security in a pure research centre; this was clear from the experience of TIFR and Bhabha Atomic Research Centre (BARC). Entry into TIFR was no hassle; whereas, entry into BARC was not that easy; it is more so at present. It is worth mentioning that currently, real estate cost in the Salt Lake area is higher than in any other part of Calcutta!

Once the cyclotron in Calcutta became operational, Ramanna felt that although Saha Institute was in a different location in the city, nuclear physicists there needed to interact with people working at the cyclotron centre and contribute to the academic atmosphere. In short, Dr. Ramanna did all that was necessary to ensure that Saha Institute was financially supported by the Department of Atomic Energy, the way TIFR was.

A governing body for SINP was constituted, and at one stage, I was designated as the member who represented the Department of Atomic Energy. What I am trying to get at is that although Saha was strongly opposed to Bhabha, Ramanna made a conscious attempt to bury the past and support good work because it was in national interest.

Ramanna helps the Young Department of Science and Technology

Born in 1928, M. G. K. Menon joined TIFR roughly at the same time as Ramanna. Menon was interested in cosmic rays, and I knew him from the time I started working with Dr. Ramanna.

In 1983, Prof. Menon was appointed Secretary of the Department of Science and Technology (DST), and he was keen to make the DST play a vital role in advancing science research in the country. He consulted Dr. Ramanna who suggested that a small group of three physicists should first prepare a note, identifying thrust areas in physics which should receive strong support.

I was one of the three selected for preparing this note, and upon receiving it, Prof. Menon arranged a three-day meeting to which about thirty or so active physicists were invited. Dr. Ramanna was requested to Chair the meeting and conduct the proceedings, and he promptly asked me to assist him; after all, I was one of the authors of the note on thrust areas in physics.

The meeting took place in Baroda, and after opening remarks by Prof. Menon, Dr. Ramanna took complete charge. He made his own opening remarks and asked me to present the note I had prepared. Several groups were then formed based on their specialisation and asked to identify thrust areas associated with their speciality. The groups met for about two days, and Dr. Ramanna moved from one group to another, to monitor what was happening.

The leader of every group was then asked to make a presentation to the entire audience, while Dr. Ramanna asked questions; my job was to listen and take notes. When group presentations were over, I showed my notes to Dr. Ramanna.

During the final session, Dr. Ramanna elegantly summarised the conclusions of the "Baroda Conclave", while Prof. Menon and his staff listened carefully and took notes. Dr. Ramanna was at his best, informal but effective, adding a touch of humour wherever possible. One of those who worked tirelessly behind the scenes was Dr. Lavakare who was once in TIFR and followed Prof. Menon to DST. I still recall Ramanna saying that the meeting was a great success because Dr. Lavakare looked after all participants with "love and care"!

For me, this was my first introduction to DST. Subsequently, I actively helped DST in many ways, and there can be no doubt that the DST played a huge role in giving research grants to thousands of researchers in various universities spread across India, and Dr. Ramanna played a key role in shaping the DST in its early days.

Raja Ramanna Brings Assistance to Many Small Institutes

Way back in the late fifties, a private institute dedicated to the pursuit of mathematics and theoretical physics was established in Madras city. It was popularly known as MatScience and there came a time when the founder retired, and the institute was in limbo. Dr. Ramanna became concerned since a couple of former members of TIFR were working in MatScience.

Ramanna then appointed a committee headed by Prof. T. Pradhan, a student of Fermi, with Prof. M. S. Raghunathan of the Faculty of Mathematics, TIFR, as one member and me as the other. The Committee visited the Institute, held detailed discussions with the staff and prepared a report recommending that the DAE bring MatScience under its umbrella and fully fund it. Dr. Ramanna accepted the report and made sure that MatScience would be duly funded. I am happy to say that since then, MatScience has done very good work. There was really no need for Ramanna to intervene, but he did because he wanted science to grow and flourish in all parts of India.

The precedence having been set, the Institute of Physics set up by Prof. Pradhan in Bhuvaneshwar soon joined the list of institutions aided by the DAE. I gather that since then the policy regarding extending aid to small scientific institutions has been substantially expanded and that the current list of institutions funded by the DAE is quite long!

Personally, I am happy to see that included in the current list is what was once called the Mehta Institute, located in Allahabad. Now renamed Harish-Chandra Research Institute, it had the distinction of having India's youngest and brightest String Theorist Asoke Sen on its faculty. Sen, now widely known, was elected FRS at a young age.

For the record, Harish Chandra graduated from Allahabad University with an M. Sc. degree in physics around 1943 or so. Harish Chandra was truly brilliant and spent some time working with Bhabha, who was then in the Indian Institute of Science, exploring ways to start his own Institute, i.e. TIFR. When the World War was over, Bhabha recommended that Harish Chandra should go to Cambridge and work with Dirac. Harish Chandra did that but found that Dirac was leaning more towards physics than pure mathematics. Harish Chandra then went to America, where he did excellent work and earned the reputation he deserved.

Ramanna is no more, but the precedent he set by granting protection to MatScience is now helping hundreds of young mathematicians and physicists in various parts of India. As for me, I am happy to have played a small part in this story.

Raja Ramanna Helps India Make a Major Decision at a Critical Moment

Early in the 1950s, President Eisenhower became concerned about the destructive capabilities of atomic energy and wanted scientists to actively explore the peaceful uses of atomic energy. Shortly after Eisenhower had suggested the "Atoms for Peace" concept, Admiral Strauss, the Chairman of the United States Atomic Energy Commission, announced that President Eisenhower wanted an international conference of scientists to be arranged soon. Strauss added that it was Eisenhower's hope that the conference would be well attended and include outstanding scientists from all over the world who would explore "the benign and peaceful uses of atomic energy".

Thereafter, things began to move fast, and in December 1954, the UN General Assembly unanimously and enthusiastically adopted a resolution which provided for the establishment of an International Atomic Energy Agency (IAEA) and for holding an international conference under the auspices of the United Nations. An Advisory Committee consisting of representatives of Brazil, Canada, France, India, USSR, United Kingdom and USA was set up to do the preparatory work for organizing an International Conference on the Peaceful Uses of Atomic Energy.

The Conference was held in August 1955 in Geneva. Dr. Homi Bhabha presided over the Conference (which clearly was a great honour for India), and Dr. Brahma Prakash, who at the time was associated with the metallurgy program of the DAE, was appointed as a Scientific Secretary. The venue of the Conference was Palais des Nations, Geneva, where excellent facilities were available for holding large multilingual conferences. The Geneva Conference lasted from the 8th to the 20th of August 1955. Thirty-eight governments submitted 1067 papers, and 1428 participants attended. The conference was wide in scope, embracing all major aspects of the peaceful applications of atomic energy. Later, the UN brought out a set of volumes containing the papers presented on various subjects. Understandably, the US and the erstwhile USSR sent the largest delegations and, as always, there were plenty of side sessions on various topics.

Let me now bring Dr. Ramanna into the picture. Since Ramanna got his Ph.D. degree for his work on nuclear fission, he was an automatic choice for the Indian delegation. After the conference was over, Bhabha was invited to the USSR, which is understandable since he presided over the Geneva Conference; besides, India was quite friendly to the USSR.

Ramanna told us many interesting stories when he came back from the Geneva Conference, but I shall not go there. All I will say is that sometime later, the UN published the conference proceedings in several volumes. Ramanna distributed those volumes to some of us and asked us to write book reviews, even though we had not even read many books! I do not remember where those reviews were published.

Looking back, I realize that Ramanna often casually assigned the job of seniors to us juniors; this was his way of making young people grow fast. Many worked under him over the years, and they developed along different lines. Ramanna deftly channelled every rookie to his or her own lane where they would be comfortable and grow suitably. It now seems to me that Ramanna wanted me in his lane!

There is a reason why I am making a big fuss about the 1955 Geneva Conference, and it relates to what I learnt from Dr. Ramanna around 1987 or so, almost accidentally I might add. At that time, I had been tasked with writing a scientific biography of C. V. Raman, whose birth centenary was to be celebrated in a grand manner in November 1988. I was then in RRC in Kalpakkam, and Dr. Ramanna came to Madras for some work. I met him there looking for some material on Raman's student K. S. Krishnan, who was there in Calcutta right next to his teacher Raman when the latter discovered in 1928 what is now called the Raman Effect.

Raman's discovery was considered so important that he was awarded the Nobel Prize in 1930, i.e. within just two years of his discovery. Much later, Bertram Brockhouse of Canada (whom many of us knew personally) described the technique of slow neutron inelastic scattering which he invented, as the "neutronic analogue" of the Raman Effect. It is noteworthy that Brockhouse won the Nobel Prize for this discovery.

With that introduction, let me now turn to Sir K. S. Krishnan. A brilliant physicist in his own right, Krishnan got his Ph.D. working under Raman and joined the Dacca University as a Reader; back then, Dacca was still a part of India! It was in Dacca that Krishnan did his pioneering work on the diamagnetism of a class of molecules. Years later, when van Vleck of America won the Nobel Prize for his work on diamagnetism, he acknowledged that the inspiration for his work came from that done by Krishnan earlier.

From Dacca, Krishnan moved to Allahabad University where he served as the Professor of Physics. From Allahabad, Krishnan went to Delhi to establish the National Physical Laboratory (NPL). Inaugurating the lab, Jawaharlal Nehru praised Sir K. S. Krishnan by saying that he, Nehru that is, would any day prefer the honour of being called the Director of NPL than as the Prime Minister! That shows Krishnan's stature. All this is fine but what has it got to do with Dr. Ramanna? That precisely is what I am coming to!

Dr. Bhabha knew K. S. Krishnan very well because like him, Krishnan too was a Fellow of the Royal Society. Being elected as FRS was always a big deal and it remains so to this day because giants such as Newton,

Maxwell, Rutherford, Eddington, Niels Bohr and Dirac had been elected FRS, not forgetting the towering Indian mathematician Srinivasan Ramanujan. It is no surprise that when the Atomic Energy Commission was first constituted, Bhabha recommended to Nehru that K. S. Krishnan should be in the Commission; Nehru agreed, immediately.

During my early days when Krishnan often visited Bombay to attend the AEC meetings, Ramanna who had a small office room in the OYC would vacate it and let Sir K. S. Krishnan use it. In later years, Ramanna introduced us to Krishnan, and we discussed neutron inelastic scattering with the big man because it was very much like the Raman Effect. What Raman did using photons we were now doing using beams of slow neutrons available from the reactor in Trombay.

Reverting to my discussions with Ramanna about K. S. Krishnan and C. V. Raman, Ramanna suddenly said to me that the world was not aware of a major contribution that Krishnan had quietly made during the Geneva Conference of August 1955. Krishnan was there as a Member of the Indian Atomic Energy Commission, more so because the Chairman of that Commission namely Homi Bhabha was presiding over a historic global meeting.

Many meetings took place on the side of the main Conference, and many agreements were also made between countries. At that time, Nehru had decided that India would remain non-aligned in the cold war between the Western powers and the Soviet Union. There in Geneva, the Americans were trying hard to convince Bhabha that light water reactors using enriched uranium as reactor fuel was the best option for India's nuclear energy program.

Bhabha had already declared that India's electric power generation HAD to come mainly from nuclear reactors. Ramanna told me that the American delegation was lobbying heavily with Bhabha, trying to persuade him to base India's nuclear energy programme on light water reactors, which used enriched uranium as fuel. For the record, I should mention that the laws of physics make it **impossible** to build a light water reactor using natural uranium as fuel. Mercifully, this is not so if heavy water is used; in other words, if heavy water is used as the moderator, it is possible to use natural uranium as fuel for the reactor.

Getting back to my story, Ramanna was worried by the heavy lobbying the Americans were engaged in and understandably so. His point was that uranium enrichment was an extremely difficult technology, far beyond India's capability at that time; this was clear from the American and British experience. Suppose one went the American way and opted for light water reactors. It would then mean that India would have to buy enriched uranium from outside for fuelling its power reactors, severely compromising India's quest for self-reliance in sensitive technologies. What if people refused to sell enriched uranium?

There was, however, an alternate option which was better suited to India. Canada had already built and successfully operated a research reactor named NRX that used natural uranium as fuel and heavy water as the moderator. The design, construction and commissioning of NRX was done by W. B. Lewis, formerly of Cambridge, who had migrated to Canada. Based on this, Canada was planning nuclear power stations based on reactors that used the natural uranium and heavy water combination.

Dr. Ramanna strongly felt that this was the way India ought to go. That being said, Ramanna was too junior to argue his case with Dr. Bhabha. Instead, he discussed his apprehensions with Sir K. S. Krishnan and pleaded that the latter should take up the matter with Bhabha. After all, Bhabha had great respect for Krishnan, besides which Krishnan carried weight with Prime Minister Nehru. Krishnan understood Ramanna's concerns, quietly met Bhabha and convinced him that India must focus on building natural uranium-fuelled, heavy water moderated reactors for its nuclear power stations.

History shows that the Americans did succeed in selling India an early version of a light water nuclear reactor, which used enriched uranium. This reactor was built in a place called Tarapur, some distance to the north of Bombay city. The conditions of sale were such that Americans would neither take away the spent fuel nor would they allow India to reprocess the spent fuel as is done elsewhere. This is precisely the sort of problem that Ramanna worried about.

The spent fuel is highly radioactive and must be kept in special ponds with a good circulation of cooling water. If cooling water is not supplied, the fuel would catch fire and tons of radioactive elements would be dumped into the atmosphere along with the smoke. This, by the way, is what happened in Fukushima in Japan some years ago, where a huge tsunami disabled the cooling pumps that circulated the water in the storage ponds. A huge fire resulted, dumping much radioactivity into the atmosphere.

In short, Ramanna foresaw clearly that India could master heavy water production far more easily than uranium enrichment. This in turn meant that India could readily use the natural uranium cum heavy water combination and have **total control over all aspects of the nuclear technology needed for building nuclear power stations**.

Following the Geneva Conference, Bhabha signed an agreement with Canada to build in Trombay a replica of their NRX reactor in Chalk River in northern Ontario. I believe this was done under an agreement known as the Colombo Plan. Dozens of Indians went to Canada to learn all they could about NRX, and they worked closely with the Canadian team that came to India to build what was called the CIR, short for Canada India Reactor. I spent more than a decade engaged in neutron scattering experiments, using neutron beams from CIR.

Meanwhile, Canada started building nuclear power stations based on what were called CANDU-type reactors. Taking help from Canada, India built a nuclear power station with two CANDU reactors at a place called Kota in the state of Rajasthan. The CANDU-type reactor is quite complicated because it allows for adding and removing fuel rods **WITHOUT** shutting down the reactor. The advantage gained is that the power station does not have to be shut down for fuel change; such shutdown is unavoidable in almost all other reactors.

Thanks to the experience gained in Kota, the Department of Atomic Energy began building two power stations in Madras, about 60 km to the south of the city and close to the shores of the Bay of Bengal. These were replicas of the reactors in Kota. Every bit of the power station, including the steam turbines and the generator, was built in India. The uranium came from India, and the tons and tons of heavy water required came from Indian Heavy Water production plants built in various parts of India. The heat exchanger, the steam turbine and the generator also were made in India.

Ramanna lived to see all this and even more. At one point, the CIR reactor renamed CIRUS, was becoming old. Trombay needed a replacement and Ramanna got Government sanction to build a new more powerful reactor. It was simply called R5 since it was the fifth research reactor to be built in Trombay. Unlike the CIR, which was a 40-megawatt reactor, R5 was planned as a 100-megawatt reactor and was essentially a more powerful version of CIR.

The R5 reactor was designed by reactor engineers working in Trombay, while the reactor control system was entirely built by the engineers of the Reactor Control Division. The reactor worked exactly as it was expected to and was later named Dhruva by Dr. Giani Zail Singh, the then President of India.

Thanks to K. S. Krishnan, Bhabha made the right decision at the right time. Unfortunately, Bhabha did not live to see what a Himalayan change it made to the country's nuclear power program; the same may be said of K. S. Krishnan also, but Ramanna sure did. Ramanna shared this history with me quite casually, when I was trying to get some material on K. S. Krishnan. When Dr. Ramanna passed away, I wrote a long article for *Physics News* titled *Dr. Ramanna as I knew him*. That was the first time this story was placed on record.

My story would not be complete if I did not make a reference to what one does with spent fuel from CANDU reactors. Bhabha decided that the spent fuel rods must not be indefinitely stored in cooling ponds. Instead, they should be reprocessed to recover plutonium and use that plutonium to build the next generation of nuclear reactors. RRC, where I worked from December 1973 to December 1987, actually built and operated a 40 MW reactor named Fast Breeder Test Reactor (FBTR) which used plutonium carbide as fuel; that Plutonium came from reactors built in India. Ramanna played a key role in the project that culminated in the construction and commissioning of FBTR. Almost all components needed for the construction of FBTR came from India, and hundreds of scientists and engineers gained from that experience.

Earlier, I made a reference to Saha and his belief that progress in atomic energy should come via the universities. Sadly, he was mistaken. World War II had transformed science. While theoretical physics could to some extent still be done using chalk and board, frontier science now calls for big projects. The classic example is CERN.

To me, CERN teaches a remarkable lesson. World War II had devastated large parts of Europe. However, when the war was over, the countries of Europe came together and decided that physics in Europe **must and should flourish again** as it had done earlier for centuries. CERN started on a small key but look where it is now! It is making even America envious.

Turning to Bhabha and Ramanna, the former was too busy with policy decisions and creating a vision for the future. Ramanna intuitively realised that he would have to play an important role in shaping that future. In that process, he enabled hundreds and hundreds of people like me to do things we never dreamt we could. More than anything else, Ramanna helped us to develop the confidence that like in the West, we too could do big things by coming together and working together.

I would like to conclude by thanking the editors of this volume for giving me a chance to pay my tribute to Dr. Raja Ramanna who mentored me in many ways, from the day I started working for him in July 1955.

Dr. G. Venkataraman, FNA, FASc, is a condensed matter physicist, with valuable contributions to neutron scattering, lattice dynamics, mechanical properties of matter, non-crystalline states, neutral networks and image processing. He obtained his Ph.D. from Mumbai University in 1966, while working at Bhabha Atomic Research Centre, Mumbai.

A few years later, he moved to Indira Gandhi Centre for Atomic Research, Kalpakkam, as Director of the Physics, Electronics and Instrumentation Group and then to Advanced Numerical Research and Analysis Group (ANURAG) of the Defence Research and Development Organisation as Director and Distinguished Scientist. After superannuation from government service, he served as the Vice Chancellor and Honorary Professor of the Sri Sathya Sai Institute of Higher Learning (Deemed University). He is the recipient of the Sir C. V. Raman Award 1974, Raman Centenary Medal 1988, Indira Gandhi Award of Indian National Science Academy 1994 and Raman Medal of the Indian Science Congress 2004. He was awarded Padma Shri in 1991 by the Government of India.

He is internationally recognised for his monumental works on the lives and works of several scientists such as H. J. Bhabha, Sir C. V. Raman, M. N. Saha and S. N. Bose and several books explaining complex but beautiful concepts of physics.

Chapter 4

Reminiscences Concerning My Academic Association with Dr. Raja Ramanna

S. S. Kapoor

Physics Group, Mumbai, India

Homi Bhabha Chair Professor, BARC, Mumbai, India

Bhabha Atomic Research Centre, Mumbai, India,

kapoorss2002@yahoo.co.in

I am happy that in the birth centenary year of Dr. Raja Ramanna – a world-renowned nuclear physicist – it is planned to bring out the book *Dr. Raja Ramanna – A Renaissance Man* with V. S. Ramamurthy and D. K. Srivastava as the editors. In their letter, the editors have nicely brought out what makes Dr. Ramanna a truly Renaissance man. In his chosen research area of fission physics, he made pioneering contributions to a deeper understanding of the nuclear fission process. It is well known that he played a leading role in the atomic energy program of the country for decades and under his leadership several important research institutions with new experimental facilities took shape in the Indian Atomic Energy Establishment. He was the architect of the atomic energy training school which has served to generate over the years highly trained manpower for the country's nuclear programs.

Around 1962 researchers from our atomic energy establishment, Tata Institute of Fundamental Research (TIFR) and the universities were pressing for their need for a medium energy particle accelerator facility in India. With the initiative of Dr. Ramanna, a meeting presided by Dr. Bhabha took place at the TIFR in which the Indian scientists from all the academic institutions made presentations to bring out the need for a medium energy cyclotron and a tandem Van de Graaff accelerator. Dr. Ramanna also asked me to make a presentation bringing out the need for a cyclotron for our research programs. In his concluding remarks, Dr. Bhabha directed that a cyclotron facility be indigenously built while a tandem Van de Graaff accelerator can be purchased later. Subsequently, it was under Dr. Ramann's leadership that a variable energy cyclotron was constructed and commissioned at Kolkata close to the campus of the Saha Institute of Nuclear Physics. The creation of the Centre for Advance Technology at Indore, now named after him as the Raja Ramanna Centre for Advance Technology (RRCAT), is another testimony of his initiative, drive and scientific leadership. In the 1980s, with his support, a 14-MV pelletron-based heavy ion accelerator was set up at TIFR as a joint BARC-TIFR facility which became a very important facility for heavy-ion-based nuclear physics research all these years.

It has been my rare privilege that I had an opportunity to get academically associated with Dr. Ramanna for several decades, starting from the time in 1959 when I joined his fission physics section to the end. In Dr. Ramanna, there was a rare combination of brilliant scientific leadership and his unassuming personality displayed in interactions with his scientific team. He had a great passion for basic research so much so that he strenuously attempted to find time to pursue his research interests, even while effectively shouldering the great responsibilities of scientific administration which go with the top positions he occupied in our Department of Atomic Energy.

Soon after the commissioning of our first reactor, Apsara in 1956, he initiated and led the basic research in nuclear fission using reactor neutron beams from this reactor and also later from the Cirus reactor in the areas which were very contemporary in those days. Under his guidance, a gridded ion chamber technique which can measure both energy and direction of emission of a charged particle was developed and the 2pi fission fragments' detection capabilities enabled us to compete with similar research programs being pursued by scientists in the USA and Europe, who had the benefit of having stronger fission sources. Thus, the research work in his group during the sixties on prompt neutrons and gamma rays and also on occasionally emitted alpha particles in fission had already put his research group in those early years on the world map in nuclear fission research.

At that time, if we could publish in competition with other laboratories abroad having better resources, it was mostly due to the ingenious experimental techniques involving gas detectors developed under his guidance and also the drive and motivation which he provided to his team. Often on a holiday, he would pick us up to take us to Trombay in his official car. He always encouraged open and critical scientific discussions without any inhibitions with respect to hierarchy. In those early days, when some eminent scientist accompanied by him, often together with Dr. Bhabha, was being taken around the various experiments at the Apsara reactor, he was always eager to project his students and would often ask me to explain what we were doing. He indeed knew how to encourage and motivate his young co-workers. His very presence was a source of great inspiration.

Dr. Ramanna started his research career at TIFR and together with holding other important positions in the DAE, he continued to be a professor at TIFR till the end. In the early sixties, he had utilized the 1MV Cascade Generator that had been set up by his group on the premises of the new TIFR campus at Colaba. With this machine, he and his group studied the slowing down of neutrons and their diffusion in water and beryllium oxide. He also led a program to study occasionally emitted long-range alpha particles in 14 MeV neutron-induced fission of U238 using this accelerator. When the 5.5 MV Van de Graaff accelerator was installed and commissioned at Trombay in 1962, he led a challenging experiment to study correlations between fragment mass-asymmetry and anisotropy in the 4 MeV neutron-induced fission of U-235, using novel ion chamber techniques. While pursuing this work, he became deeply interested in the mechanism of the fission process responsible for the asymmetric fragment mass distribution. He was very original in his thoughts and ideas. He tried to understand this complex mechanism by choosing a line away from the beaten track. In the sixties, while reading Prof. S. Chandrasekhar's paper on random walk, he applied those ideas to have a completely original approach to understanding mass distributions in nuclear fission based on the Markovian process of nucleon exchanges between the two nascent nuclei evolving from saddle-to-scission in fission. This approach to understanding the mechanism of nuclear fission was totally unconventional and new. Later, many details of this stochastic model of fission were worked out by him and co-workers and it was seen that the broad features of mass distribution in fission can be accounted for by this model. When he wrote to the famous nuclear physicist Prof. Eugene P. Wigner about this work, Prof. Wigner replied with the comment that we should also see this nucleon exchange process in heavy-ion reactions. Some years later when several heavy-ion accelerators became operational in laboratories around the world, this process of nucleon exchanges between the two colliding heavy ions at medium energies indeed became an important subject of study around the world and the nucleon exchanges between two nuclei in contact was experimentally seen in the heavy ion deep inelastic collisions which supported his ideas. With this development, he developed a keen interest also in heavy ion-based nuclear physics research and we would often discuss the latest in heavy ion reaction studies, particularly deep inelastic collisions.

In those days when I joined his fission physics section, Dr. Ramanna's office was at the Old Yacht Club (OYC) near the Gateway of India. A library that had technical books and journals was also located at the OYC building. When I visited the library at OYC, I would also often go and meet him in his office room to report on the experiments and for general discussions in the areas of fission physics and nuclear physics. While he was holding the top scientific position and I was just a fresher, he always encouraged me to come and see him for scientific discussions. Also, he used to frequently come to the Apsara reactor building to monitor the progress of our ongoing experiments and for scientific discussions. He was there certainly on Tuesdays, as on this day, the Trombay Council meetings were being held in the Apsara building conference room on the first floor. Occasionally, we would also see Dr. Bhabha there on the Trombay Council meeting day. I remember that once Dr. Bhabha accompanied by Dr. Ramanna came to our experimental setup at the Apsara reactor. Here, Dr. Ramanna described to him our experiment during which I had a rare opportunity to also speak to Dr. Bhabha. When Dr. Ramanna came to our experimental setup at the Apsara reactor after having discussions on the experiment, he would often accompany us to the south site canteen for tea. On many occasions, while returning home after work, he used to offer me a ride in his car, where we would continue to have our scientific discussions regarding our experiments. There was one incident which made a deep impression on me. I was staying in a small flat at the Kenilworth building on Pedder Road while Dr. Ramanna was staying at the Anand Bhavan building on Warden Road. One Sunday afternoon, sometime in 1962, my doorbell rang, and I saw Dr. Ramanna standing at the door. He said he had come to discuss some points in the paper we were writing. This incident showed his unassuming personality and that he did not allow the difference in our ranks to come in the way of our free and frank scientific discussions. With the publication of some important research work by Dr. Ramanna and his team as given in the following, his research group became an important centre of neutron-based studies and fission research (Fig. 1).

Fig. 1. The Fission Physics Group of BARC: Second row, third from the left: Dr. V. S. Ramamurthy and fifth from the left, the author. Seated first row, second from left: Dr. A. K. Mohanty, the present Chairman of Atomic Energy Commission.

In 1964, I received my Ph.D. degree from Bombay University with Dr. Ramanna as my Ph.D. guide, and around that time, I left for the USA to work at the Lawrence Laboratory at Berkley. When I was there, I was pleasantly surprised that Dr. Ramanna had arranged with the IAEA for my participation at the first International Fission Conference being held at Salzburg in March 1965, which he was also attending. I remember many detailed discussions on fission physics he had there with other experts in the field.

In India, the first Peaceful Nuclear Explosion (PNE) was carried out in 1974 when Dr. Ramanna was the Director of BARC and his pivotal role in this work is common knowledge now. Around 1978, he moved to Delhi as Scientific Advisor to the Minister of Defence and as Chief of DRDO. In the few years that he worked there, DRDO saw important reforms in its functioning. Around 1981, he returned to his parent organization as Director of BARC and later around 1984 as Chairman of the Atomic Energy Commission. When he was in Delhi, my academic interaction with him was not that frequent. I would meet him there only during some of my tours to Delhi and would discuss new developments in areas of mutual interest. Later, when he moved to Bengaluru as Director of NIAS, he continued his interest in fission physics, heavy ion reactions and the newly emerging topic of super-heavy nuclei. When he was visiting Trombay, he would find time to meet us and have discussions on the latest in these areas.

Dr. Ramanna was academically active till the end. About two weeks before the end came, I remember receiving a phone call from him when we discussed new developments in the above areas of interest, and he asked me to send him some transparencies on these topics as he was planning to give a lecture to some students to convey to their excitement of basic research.

In the biographical memoirs of Fellows of the Indian National Science Academy, Vol. 28, Professor B. V. Sreekanthan, former Director of TIFR, has written on Dr. Raja Ramanna in more detail and I quote in the following a few selected lines from his write-up:

> *Raja Ramanna was a rare combination of a scientist, technologist, administrator, philosopher, musician and musicologist who played a dominant role for over 5 decades in the advancement of science and technology in the post-independence era of India. His genial temperament, very positive and helpful attitude, and extreme transparency in dealings made everyone comfortable to interact with him both at professional and personal levels. He was a Colossus in terms of achievements, but very simple and modest in his daily life.*

In the end, I again pay my respectful homage and tributes to *Dr. Raja Ramanna: A Renaissance Man*. I am adding references to his most celebrated papers below.

References

1. R. Ramanna, G. S. Mani, P. K. Iyengar, S. B. D. Iyengar & B. V. Joshi, *Proc. Int. Conf. on Peaceful Uses of Atomic Energy (IAEA)*, (Geneva) 1955 P/872, 24.
2. R. Ramanna & N. Sarma, *Proc. Symposium on Inelastic Scattering of Neutrons in Solids and Liquids (IAEA)*, (Vienna) 1961, 631.

3. R. Ramanna, R. Choudhury, S. S. Kapoor, K. Mikke, S. R. S. Murthy and P. N. Rama Rao, The angular correlation of prompt neutrons and fission fragments in the thermal neutron fission of ^{235}U, *Nucl. Phys.* 25, 136 (1961).

4. R. Ramanna, K. G. Nair and S. S. Kapoor, Emission of alpha particles in the fission process, *Phys. Rev.* 129, 1350 (1963).

5. S. S. Kapoor, R. Ramanna and P. N. Rama Rao, Emission of prompt neutrons in the thermal neutron fission of ^{235}U, *Phys. Rev.* 131, 283 (1963).

6. S. S. Kapoor and R. Ramanna, Emission of prompt gamma rays in the thermal neutron fission of ^{235}U, *Phys. Rev.* 133, B598 (1964).

7. S. S. Kapoor, D. M. Nadkarni, R. Ramanna and P. N. Rama Rao, Kinetic energy distributions and the correlation of Anisotropy and asymmetry in the 4 MeV neutron induced fission of ^{235}U, *Phys. Rev.* 137, B511 (1965).

8. R. Ramanna, Mass distribution in fission and the theory of random flights, *Phys. Lett.* 10, 321 (1964).

9. R. Ramanna, R. Subramanian and R. N. Aiyer, Nuclear fission as a Markov process, *Nucl. Phys.* 67, 529 (1965).

10. R. Ramanna and V. S. Ramamurthy, Mass and charge distribution in fission, Proceedings of Second IAEA Symposium on Physics and Chemistry of Fission, p. 51, IAEA, Vienna (1969).

11. M. Prakash, V. S. Ramamurthy, and S. S. Kapoor, Studies in the statistical theory of nuclear fission and explanations of fragment mass asymmetry in terms of nucleon exchange mechanism, *Proceedings of IAEA Symposium on Physics and Chemistry of Fission Julich*, p. 353, IAEA, Vienna (1979).

12. S. S. Kapoor, V. S. Ramamurthy and R. Ramanna, Fusion-Fission angular distributions: A new probe of fast fission fractionation in nucleus-nucleus collisions, *Pramana-J. Phys.* 22, 275 (1984).

Dr. Shyam Sunder Kapoor, FNA, FASc, FNASc, FIoP, joined the Atomic Energy Establishment [now Bhabha Atomic Research Centre (BARC)], Trombay, in 1959, after his M.Sc. (1958) in Physics from Agra University. He obtained his Ph.D. degree (1963) from Bombay University in Nuclear Physics under the guidance of Raja Ramanna. He carried out post-doctoral research (1964–1966) in nuclear fission at Lawrence Berkeley Laboratory, Berkeley, and was a Visiting Scientist (1980–1981) at the Physikalische Institute, University of Heidelberg. At BARC, he rose to become Head of Nuclear Physics Division and also Project Director of the Pelletron Accelerator facility at Tata Institute of Fundamental Research and Director of the Physics Group and Electronics and Instrumentation Group at BARC (1990–2000). He was DAE-Homi Bhabha Chair Professor (2000–2005). Dr. Kapoor is well known for his pioneering research contributions in the areas of nuclear fission, nuclear shell effects, nuclear dynamics of heavy-ion fusion-fission reactions, super-heavy nuclei, nuclear radiation detectors and particle accelerators and is a recipient of several prestigious awards and prizes.

https://doi.org/10.1142/9789819814435_0005

Chapter 5

Dr. Raja Ramanna and His Legacies

V. S. Ramamurthy

National Institute of Advanced Studies, Bengaluru, India

vsramamurthy@gmail.com

Dr. Raja Ramanna occupies a very special place in the history of nuclear research in India. After obtaining his Ph.D. degree in Physics from Kings College, London, Ramanna joined the Tata Institute of Fundamental Research, Mumbai, in 1949. In those years, nuclear research was still under a veil of secrecy because of its demonstrated weapon potential. It was only in 1955 that the first International Conference on Peaceful Uses of Atomic Energy was held in Geneva and nuclear research was substantially declassified. It was also the year in which Dr. Homi Jehangir Bhabha conceptualized India's first nuclear research reactor, Apsara. The reactor was designed and built soon after and went critical in 1956. While the reactor was used primarily to produce radioisotopes for diverse applications, Ramanna also saw an opportunity to carry out basic research in fission physics. The Trombay Fission Group, under the guidance of Ramanna, was one of the few groups outside the Western world to undertake full-time research on nuclear fission that early. It was this group that I joined after completing my training at the Atomic Energy Establishment Trombay (AEET, now renamed as Bhabha Atomic Research Centre) Training School in 1964.

Those were hard days. Carrying out globally competitive nuclear research was indeed tough. The resources were limited. Foreign exchange was scarce. Nuclear instruments were not available off-the-shelf. You had to design and fabricate your own detectors, pulse analysis systems, data recording systems, etc. Travel and communications were often unaffordable. Intellectual isolation was indeed a reality with very few opportunities to interact with scientists from across the world. There was always a lurking fear – whether one could do competitive research with all these constraints. Dr. Ramanna always used to say, "Develop your expertise to a level when you have full faith in it. Limited resources, facilities and manpower are unavoidable constraints in globally competitive research but R&D being primarily a human-centric activity, it is important to unshackle your mind and do your best. Respect your peers but do not be intimidated by them. In matters of scientific research probing the unknown, peers are as ignorant as you are".

We learnt a lot from Ramanna. One thought that is uppermost in my mind is that Ramanna was never a "safe" science man. What do I mean? How do we all select the topics for our research? The topic should be of current interest, preferably international interest. How else do you publish in journals of repute, get invited to meetings and conferences, build your citation index, etc.? Ramanna never believed in this safe approach. He used to ask, "Is there an unanswered question, can you look at it differently?" For example, when I joined his group in 1964, he was looking at the asymmetric mass distribution in low-energy fission of heavy nuclei that had defied an explanation based on the known nuclear physics at that time. Ramanna was visualizing a rapid exchange of nucleons between two nascent fragments, a nucleon diffusion between the two nascent fragments, just prior to scission. He even estimated a diffusion coefficient for nucleon exchange between the nascent fragments. It is not surprising that hard-core nuclear physicists, used to rigorous quantum mechanical approaches, hesitated to accept fission research as an integral part of nuclear physics research.

I recall an interesting encounter with a former nuclear theorist, Prof. Manoj Banerjee, of Saha Institute of Nuclear Physics, Calcutta. The 1965 DAE Nuclear and Solid-State Physics Symposium was being held at Saha Institute of Nuclear Physics, Calcutta. I distinctly recall the heated discussions between Dr. Ramanna and Prof. Manoj Banerjee on the nucleon exchange model. Prof. Banerjee, being a hard-core nuclear theorist, was highly sceptical of the nucleon exchange model. It is well known that nuclei are quantum many-body systems with very complicated interparticle interactions. It is not surprising that the initial theoretical developments on large amplitude collective motions as in fission dynamics were predominantly based on phenomenological models. Although appropriate nuclear many-body theories were being formulated, these were hardly adopted in practical applications because of computational limitations. After several prolonged discussions between the two stalwarts, Prof. Banerjee said to Ramanna during the evening dinner cruise, "Ramanna, we can keep on arguing on this without coming to any definite conclusion. The only way to bring this to an end is to build an accelerator of sufficient energy to bring two nuclei within the range of nuclear forces and see whether nucleon exchanges do actually take place". It is heartening to note that within a decade, nuclear accelerators were built to bring two heavy nuclei within the range of the nuclear force and observed nucleon exchanges in the so-called deep inelastic collisions. The diffusion coefficients derived from the experimental data were indeed quite close to what Ramanna had conjectured a decade earlier.

Rapid washing out of nuclear shell effects with the excitation energy was another significant contribution from the Trombay Fission Group. The discovery of fission isomers in the mid-seventies led to the concept of macroscopic–microscopic approach to nuclear masses, deformed shells and double-humped fission barriers. We recognized an anomaly almost immediately. While the new results on fission fragment angular distributions in near barrier fission of several actinide nuclei were consistent with the new concept of double-humped fission barriers, it was clearly at variance with earlier results on fission fragment angular distributions in alpha-induced fission reactions that were consistent with the liquid drop model. This led us to postulate a rapid vanishing of nuclear shell effects with excitation energy and, additionally, highlight the need to produce cool compound nuclei in the efforts to produce super heavy elements through energetic heavy ion collisions.

Not being a "Safe Science" man, Ramanna always dared to differ. If you went to him and said, "I have carried out this investigation and my results are in good agreement with all previous investigations", he would say, "Congratulations. You have done a good job, but this is not the problem where you spend more time". If on the other hand, you said, "I have carried out this measurement and I have a problem reconciling my results with other existing measurements", he would say, "Very good, double check your results. If the discrepancy persists, this is where you should concentrate". In Ramanna's view, discrepancies and anomalies were possible precursors of new information. Chasing anomalies was the working principle of Ramanna. This we did with quite a bit of success.

Chasing anomalies paid us rich dividends in yet another class of heavy ion-induced fission-like reactions. Based on anomalies noted in a number of heavy-ion induced fusion–fission reactions, we postulated a new mode of fission–pre-equilibrium fission, a lethargic relaxation of the reaction plane and a discontinuity in the fusion dynamics across the Businaro–Gallone point in the entrance channel. Our conjecture has been validated by many subsequent experiments.

The Peaceful Nuclear Experiment (PNE) in 1974 brought out another facet of Dr. Ramanna's expertise. As you are all aware, Dr. Ramanna was the main architect of this mega project. I had only a small role to play in this project as a member of the design team, but I had an opportunity to watch from close quarters Dr. Ramanna's project management capabilities. Mobilizing a large multidisciplinary team drawn from several institutions across the country and successfully completing the assignment, all the while maintaining a high level of secrecy, was no mean achievement.

I moved out of the Bhabha Atomic Research Centre in 1989 and ended up in New Delhi in 1995 as Secretary of Department of Science and Technology. DST is a major funding agency for scientific research in the country and I can assure you that no proposal will go through the DST Peer Review system until one follows the "scientific method" – a method based on well-defined objectives, systematic experiments and analysis. Of course, we know that the so-called "accidental" discoveries do not follow this "think-straight" method. However, the present scientific system looks at "accidental" discoveries as exceptions rather than the rule. While the new system has "delivered" by way of new scientific discoveries, technologies, new products and new services, there are concerns. The system obviously delegates individual scientists driven by their curiosity alone to the background. The ubiquitous peer review system for research funding is a clear disincentive for out-of-box thinking. The research priorities are likely to be distorted by funding agencies. The neutrality of science and scientists could also be in question. It is in this context that Ramanna's message to all of us becomes highly relevant.

This discussion would not be complete without talking of Dr. Ramanna's charm and humour, and his generosity to put people around him at ease. One of those intangible qualities that Dr. Ramanna had was that he effortlessly brought out the best in the people around him, and they felt privileged to be working with him and knew that their efforts would not be in vain.

Dr. Ramanna's humour was never at the cost of someone and always had the gentle touch of the soothing sunlight on a cold day. It also invariably diffused tense situations and put people at ease.

The Trombay Fission Group is perhaps one of the few groups in the world having a sustained program of research on fission and fission-like reactions over several decades. The Trombay Fission Group exists even today and is contributing significantly to the area of fission induced by reactor neutrons and fission-like reactions induced by medium energy heavy ions. We owe this to Dr. Ramanna.

Yet another initiative in which Ramanna played a crucial role was the organisation of the annual DAE Symposia on Nuclear Physics, Solid State Physics and High Energy Physics.

Based on a suggestion by Dr. Bhabha to organise a symposium where all the practitioners of Nuclear Physics, Solid State Physics, High Energy Physics and Nuclear Instrumentation could meet regularly, present their findings and discuss their results, Ramanna was instrumental in organising the annual DAE symposia on these topics. The first two were held in Bombay in 1957 and 1958. In the early days, the subjects which were discussed under the heading High Energy Physics now were also covered by the DAE NPSSP Symposium. It branched out to have a separate DAE Symposium on High Energy Physics. As the total number of participants increased over the years to exceed 500–600, it was split into Symposium on Nuclear Physics and Symposium on Solid State Physics. The 68th edition of the Nuclear Physics Symposium was held at Indian Institute of Technology, Roorkee, in December 2024 with over 600 participants – about 40% of whom were women! The DAE Symposium on Solid State Physics now routinely hosts about 1000 participants.

Right from the beginning, there was an effort to hold these yearly symposia in different universities and institutes in different parts of the country. The symposium led to the explosive growth of these subject areas across the country and so also the demands for additional facilities such as accelerators and detectors – for which the proposals were discussed at various symposia and other satellite meetings.

For years, Dr. Ramanna delivered the concluding remarks at these symposia. These used to be the high point of the meetings as he identified the threads and possible future directions of the presentations and picked out young researchers for their outstanding work. His equanimity and even-handed enthusiastic support for young researchers, and his humour, used to be in full flow during these meetings – where students surrounded him and hung on to every word that he said.

Dr. Ramanna used his skills as an organiser and motivator to build consensus and later get clearances for the installation of the Van de Graaff generator at Trombay; Variable Energy Cyclotron and later Super Conducting Cyclotron at Kolkata; the Pelletron Heavy Ion Facility at TIFR; and the Synchrotron Radiation Facilities, INDUS-1 and INDUS-2 at Indore. This process of consensus building initiated by Dr. Ramanna has pervaded the science atmosphere and was again followed for various international science collaborations, e.g. with CERN, FAIR, ITER, LIGO, SKA and TMT.

Getting people from all parts of the country and all institutions and universities of the country to work together was evident in everything that Dr. Ramanna decided to get done – and this was demonstrated time and again during his life.

He led people with humour, wit, affection and understanding and earned their respect due to his brilliance and commitment. He showed the way out of difficult situations but did not show it. He also never hurt others! Dr. Ramanna rejoiced in the success of his colleagues.

Dr. Ramanna and the AEET Training School

It is well known that Dr. Bhabha in his famous letter to the Tata Trust, while making a case for a modern research institute, also talked of getting trained manpower within the country, to lay the foundations of modern science, technology and industry. Once the Atomic Energy Establishment (later renamed as the Bhabha Atomic Research Centre) took shape, Dr. Bhabha entrusted Dr. Ramanna with the responsibility of setting up of the Training School, with the objective of imparting specialized education and training in the area of nuclear science and engineering to fresh graduates selected on the basis of their academic performance and a rigorous interview (later, through a written test and then interview). The first batch started in 1957 and operated from rented classrooms and hostels in the beginning. Dr. Ramanna designed courses and oversaw teaching, training, experimental facilities and even the day-to-day running of the Training School. It is said that he knew every trainee by name and kept track of their professional career for decades. The success of this endeavour is seen by the exponential expansion of the activities of the DAE in almost all branches of science and engineering, for which the entire necessary manpower has been trained there.

Dr. Ramanna's messages are as relevant today as they were decades ago when he was our guru.

Prof. V. S. Ramamurthy is a well-known Indian nuclear scientist with a broad range of contributions from basic research to science administration. Prof. Ramamurthy started his career at Bhabha Atomic Research Centre, Mumbai, in the year 1963. He has made important research contributions, both experimental and theoretical, in many areas of nuclear fission and heavy ion reaction mechanisms, statistical and thermodynamic properties of nuclei, physics of atomic and molecular clusters and low-energy accelerator applications. During the period 1995–2006, Prof. Ramamurthy was fully involved in science promotion in India as Secretary to the Government of India, Department of Science & Technology (DST), New Delhi.

He was also the Chairman of the IAEA Standing Advisory Group on Nuclear Applications for nearly a decade. After retirement from government service, Prof. Ramamurthy, in addition to continuing research in Nuclear Physics at the Inter-University Accelerator Centre, New Delhi, has also been actively involved in human resource development in all aspects of nuclear research and applications. Prof. Ramamurthy is also a Chairman, Recruitment and Assessment Board, Council of Scientific and Industrial Research and a Member of National Security Advisory Board. In recognition of his services to the growth of science and technology in the country, Prof. Ramamurthy was awarded one of the top civilian awards of the country, the Padma Bhushan, by the Government of India in 2005.

Prof. Ramamurthy was the Director of the National Institute of Advanced Studies from 2009 to 2014 and is now an Emeritus Professor there.

https://doi.org/10.1142/9789819814435_0006

Chapter 6

Nurturing the Training School and Supporting the Homi Bhabha National Institute

R. B. Grover

Atomic Energy Commission, Chairman of the Board of Research in Nuclear Sciences, Mumbai, India

Homi Bhabha National Institute, Mumbai, India

rbgrover@hbni.ac.in

Introduction

The Department of Atomic Energy (DAE) was set up on 3 August 1954, and in November 1954, a conference on the DAE for Peaceful Purposes in India was organised in New Delhi. At the conference, Homi Bhabha presented a general plan for the development of nuclear energy in India. The plan advocated the pursuit of a closed fuel cycle [1]. In August 1955, at an international conference in Geneva, Homi Bhabha presented a paper to delineate the role of nuclear energy in India's energy mix [2]. Taken together, these two papers inform the reader of the magnificent plan of Homi Bhabha for the growth of nuclear power and fuel cycle facilities in India. Such a magnificent plan can be implemented only by trained manpower, but Bhabha's attempts to modernise the university syllabus were unsuccessful. Therefore, he decided to set up a school to provide training in nuclear science and engineering to young graduates from universities in India [3]. This method of building up staff for the atomic energy programme did not drain away senior persons from the universities and provided training and employment to young graduates [4].

The first atomic research reactor "Apsara" became operational in 1956. This was a landmark achievement. Soon after, in 1957, the Training School was established. It is the backbone of the atomic energy programme in India. Raja Ramanna was assigned the task of implementing this training programme. His nurturing of the programme in its formative years provided it with a strong base, and as a result, the programme continues to flourish.

On 3 March 2018, a grand programme was organised by Bhabha Atomic Research Centre (BARC) to celebrate the diamond jubilee of the Training School. Alumni of the first batch were invited, and I had the opportunity to speak with many of them. Before the function, the Training School contacted several past alumni to know their impressions, and these were included in the brochure published for the occasion [5]. Some of those impressions are reproduced here to bring out the importance of the Training School and the role of Raja Ramanna:

- "Training School, nurtured initially by Raja Ramanna, has completely justified the faith reposed in it. It has played a major role in making the country self-dependent in all aspects of Atomic Energy and allied fields of S&T and, in addition, to some other national efforts. BARC-Training School has made DAE proud; it has made India proud" B. A. Dasannacharya, 1st Batch.
- "Over the years, the training programme has proved to be the melting pot for moulding scientists and engineers with technical capability comparable to international standards. The credit for this goes to the well-structured programme of training which remained alive to the dynamic needs of the rapid developments in this sector, the rigourous selection process of the trainees, the commitment of the adjunct teaching faculty and the continued involvement and support from the top management" S. K. Mehta, 1st Batch.
- "Realizing that the quality of the manpower engaged in its R&D programme is the most important single factor for its success, DAE had the foresight to start the training school 60 years back when it was in the early stages of its scientific activities. Looking back at these sixty years, one can be justly proud of the Training School which has been

playing the most pivotal and crucial role in the success of the R&D programme of the DAE" S. S. Kapoor, 2nd Batch.

- "Indian Scientists were trained in various Atomic Energy related subjects at the BARC Training School, which were usually not included in the then-existing syllabi of Indian Universities. The effort has paid off very rich dividends as can be seen from the fact that at later stages, the Training School has provided top-class leaders for most of the responsibilities of the entire spectrum of activities of the Department of Atomic Energy in all facets of the Nuclear Cycle" Jai Pal Mittal, 3rd Batch.

- "Thanks to one year in the training school, we have a large cadre with a shared vision in the arena of atomic energy for the country and the passion to implement and carry it forward. Induction through training school and career progression through merit promotion scheme are the key elements of excellence in the organisation including leadership development" Anil Kakodkar, 7th Batch.

- "Looking back to the role played by the BARC Training School over the decades, I feel indeed proud to recognise the fact that it has been the backbone of our success in establishing the complex Nuclear Science and Technology ecosystem in the country to improve the quality of life in our society" B. Bhattacharjee, 9th Batch.

- "During the last sixty years, thousands of Engineers and Scientists have been trained in nuclear science and engineering by the Training School. The institute has also continuously upgraded the training programme to meet the current requirements. Expert faculty is drawn from various units of DAE to provide holistic training across a wide spectrum of domains. The trained engineers and scientists have contributed immensely to the phenomenal growth of DAE and making the country attain self-reliance in nuclear science and technology" Chandrakant Pithawa, 17th Batch.

The Training School

Interactions with the alumni of the first batch informed me about the difficult beginning because of a make-shift hostel and the absence of a dedicated campus or building. Classes were held in different places in Bombay (now Mumbai) for scientists. Engineers received a part of the training at the Indian Institute of Science, Bangalore (now Bengaluru).

This has now changed and over the years, the Training School has evolved into a graduate school of international repute having a building, a hostel and a well-designed syllabus. Unlike universities, where the decision-making process proceeds at a glacial speed, the syllabus at the Training School is continuously updated. Keeping the syllabus up to date was a motivating factor for establishing the school independent of universities.

Initially, the selection for the school was via interviews conducted by scientists and engineers. The increasing number of applicants made the interview process burdensome. Therefore, an all-India written test was added to screen the candidates to limit the number interviewed. Interviews of those selected last as long as one hour, the questions are strictly academic and to facilitate a good conversation, an option to speak in a language other than English is offered to those who so desire. The selection is very competitive.

Details on selection percentage, disciplines and faculty can be seen elsewhere [6], but some important aspects related to the evaluation of students and academic load are reproduced here.

"Examinations are conducted almost throughout the year. A unique feature of the examination system is the end-semester viva voce. Rather than conducting viva voce for a subject, it is conducted for all the subjects taught in a semester. The problems the students will be solving as employees will not belong to a discipline and by taking viva voce for all disciplines together, the attempt is to make a judgement about the problem-solving ability of the students.

While in the Training School, trainees find the academic load very high and more diverse than required, but after graduation, they invariably agree that the intense training programme with exposure to a wide variety of subjects had equipped them with a wholesome background for the tasks assigned to them".

Over the years, the Training School was nurtured by several stalwarts (Fig. 1) and moved to a beautiful building located in picturesque surroundings on the South Site (Fig. 2) of BARC in 1970. Due to certain logistic issues, it was moved to its present building (Fig. 3) in Anushakti Nagar in 2009. The author belongs to the batch of the Training School that was the first to study in the building at the South Site, BARC, and has the privilege to have his office in the present building.

The Training School, nurtured by Raja Ramanna in its formative years, has been growing from strength to strength and has been the

Our Leaders

Dr. K.K. Damodaran
1957-1981

Dr. M.P. Navalkar
1981-1989

Dr. U.C. Mishra
1989-1993

Dr. R. Subramanian
1993-1996

Dr. H.R. Siddiqui
1996-1999

Dr. S.P. Garg
1999-2005

Dr. R. R. Puri
2005-2012

Dr. B.K. Dutta
2012-2015

Dr. A.P. Tiwari
2016-2021

Dr. D.M. Gaitonde
2021-2022

Dr. A.K. Dureja
2023 till present

Fig. 1. Leaders of the Training School.

backbone of the atomic energy programme. The need for innovations germinated the idea of establishing a university-level institution and the experience in running the Training School played an indirect role in establishing that as described in the following section.

Fig. 2. The erstwhile Training School Complex (1970–2008), South Site, Trombay, Mumbai.

Fig. 3. The Training School and the Central Office of HBNI, Anushakti Nagar, Mumbai.

Homi Bhabha National Institute

Innovation needs input at all levels: research, technology development and technology deployment. DAE has set up a wide network of institutions to implement all these levels. With the availability of uranium from the international market after recent changes in the regime governing international civil nuclear trade [7], it is well poised to increase installed nuclear power capacity based on available and innovative technologies.

DAE has two types of institutions: those which are directly managed by the Central Government such as BARC and Indira Gandhi Centre for Atomic Research (IGCAR) and those that are managed by a trust or a society and function as grant-in-aid institutions such as Tata Institute of Fundamental Research (TIFR) and Saha Institute of Nuclear Physics (SINP). While the focus in institutions managed by the Central Government is more on technology development, the focus in grant-in-aid institutions is more on research. To nurture innovation leading to deployable technologies, one needs both research and technology development and also intense collaborations between all DAE institutions. This called for devising a framework which can accelerate innovation and nudge all towards cooperation.

While all DAE institutions have been running academic programmes for awarding academic degrees, they were affiliated with various universities in the country, an arrangement not considered satisfactory. The functioning of the management in the university system frustrates any attempt to improve the rigour of academic research and innovate syllabi. Doctoral research has twin objectives: training a student in research and doing research. By choosing topics related to the mission of the Department for Doctoral Research, a lot can be achieved as a thesis documents four man-years of research by a bright young person who normally works very hard. This motivated the Department to set up a university that can help in the growth of doctoral research relevant to the mission of DAE, provide for joint guidance of students from within and across DAE institutions, make course work and general comprehensive examination an integral part of a Ph.D. and improve the quality of research that goes in a thesis. A proposal to establish a DAE university was mooted and endorsed by various bodies within the Department including the DAE Science Research Council. DAE Science Research Council (DAESRC) consisting of eminent scientists was constituted in 2001 to advise DAE in identifying thrust areas to be taken up for research, upgrade the level of ongoing research and other important issues. The Council was chaired by Raja Ramanna, and I was its secretary. Establishing Homi Bhabha National Institute (HBNI) was one of the important issues discussed by the Council and Raja Ramanna endorsed it. There was a debate on whether to call it a university or an institute; finally, we decided on it being an Institute.

HBNI was accredited as a deemed-to-be-university in June 2005 and has been steadily progressing, but on the way, there was one major challenge. A review committee was constituted by the Ministry of

Human Resource Development (now Ministry of Education) to examine all deemed to be universities. This committee devised a set of parameters for examination and one parameter was "The idea of a university". Under this parameter, HBNI was given zero marks and overall HBNI was placed under the category "B". This placed HBNI in a very disadvantageous position and needed to be changed. Members of the Review Committee were all eminent educationists and had certain notions about a university. However, all of them were open to dialogue. To have that dialogue, I had to read all about debates on the topic of the idea of a university and concluded that several ideas of a university co-exist around the world and even within a country. Based on this knowledge, we presented our case once again to the review committee, and HBNI was upgraded to category "A". I still keep reading on the topic and recently wrote about it [8].

HBNI nurtures the academic programmes of the following institutions:

- Bhabha Atomic Research Centre (BARC), Mumbai
- Indira Gandhi Centre for Atomic Research (IGCAR), Kalpakkam
- Raja Ramanna Research Centre for Advanced Technology (RRCAT), Indore
- Variable Energy Cyclotron Centre (VECC), Kolkata
- Saha Institute of Nuclear Physics (SINP), Kolkata
- Institute for Plasma Research (IPR), Gandhinagar
- Institute of Physics (IoP), Bhubaneswar
- Harish-Chandra Research Institute (HRI), Prayagraj
- Tata Memorial Centre (TMC), Mumbai
- Institute of Mathematical Science (IMSc), Chennai
- National Institute of Science Education and Research (NISER), Bhubaneswar
- Homi Bhabha Cancer Hospital & Mahamana Pandit Madan Mohan Malviya Cancer Center (HBCH & MPMMCC), Varanasi.

NISER became an off-campus center in 2016 and HBCH & MPMMCC in 2024. All others are Constituent Institutions.

With the setting up of HBNI, the Training School programme has been converted into coursework for an M. Tech. from HBNI. Students who complete a project after the coursework become eligible for an M. Tech. The post-graduate programmes in medicine (mainly at the Tata Memorial Center and some at BARC) have grown significantly (Fig. 4). The number of students getting a Ph.D. from HBNI has steadily increased (Fig. 5).

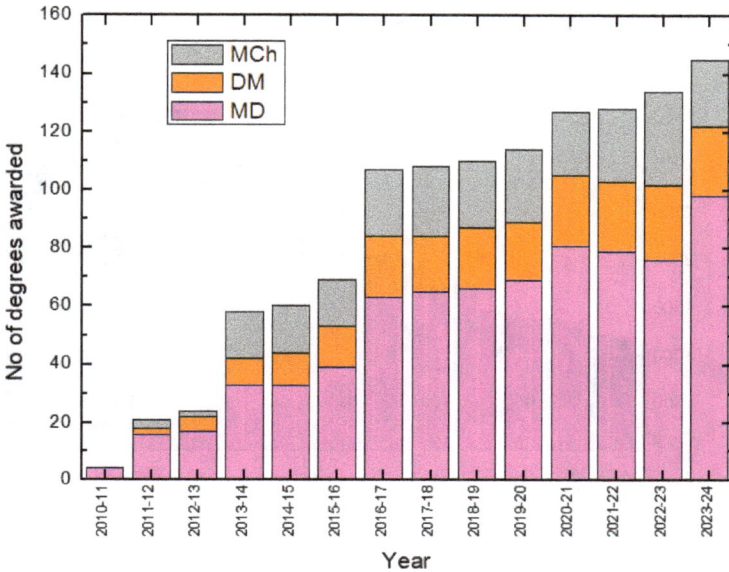

Fig. 4. Year-wise MD/DM/MCh results declared.

Fig. 5. Year-wise Ph.D. results declared.

The presence of industrial units within DAE itself gives HBNI students an inherent advantage. All these industrial establishments were started by DAE to deploy the products and processes developed by its R&D centres. Through a choice of research problems highly relevant to these organizations, the students get an opportunity to contribute to the growth of the nuclear industry. At the same time, publications from HBNI have been steadily increasing (Fig. 6).

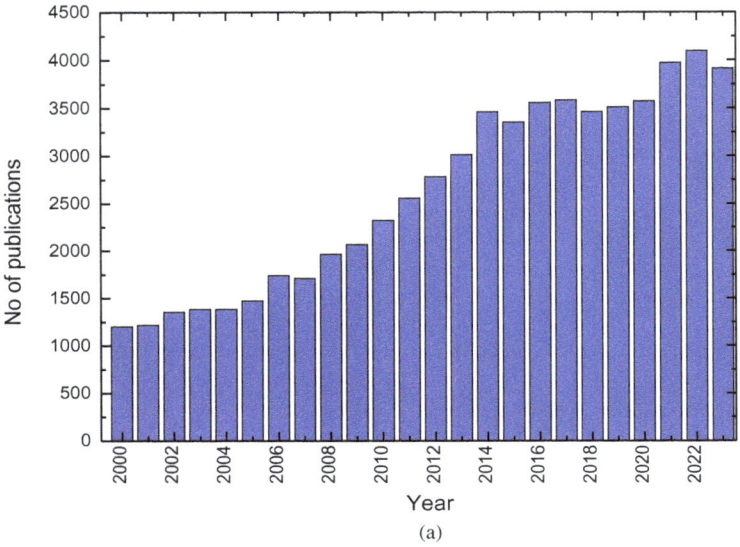

(a)

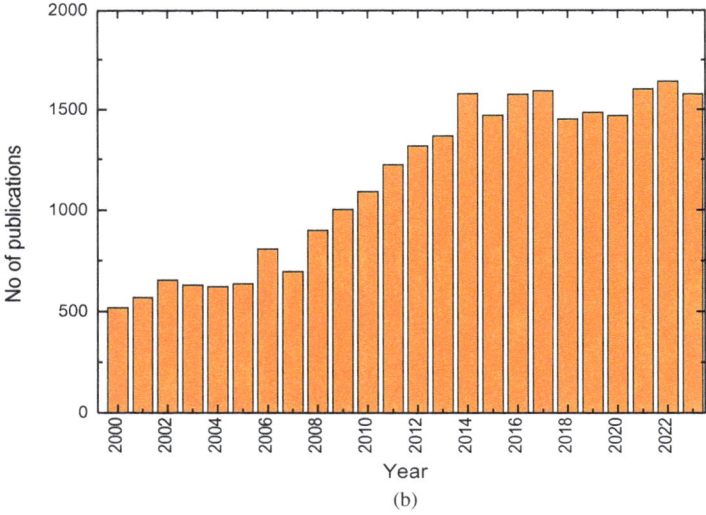

(b)

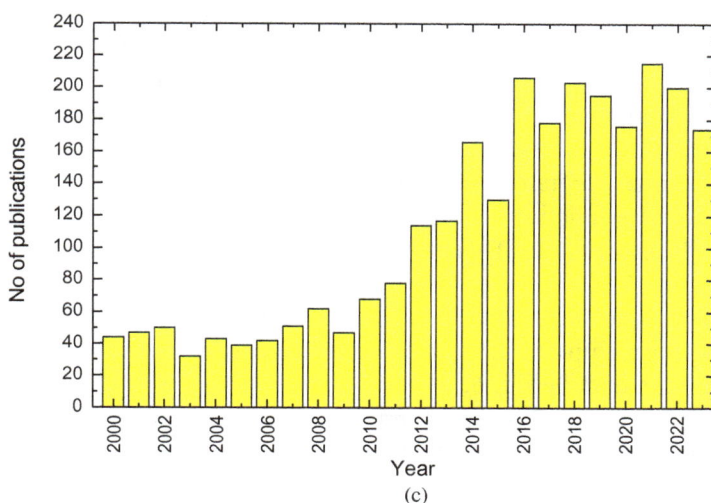

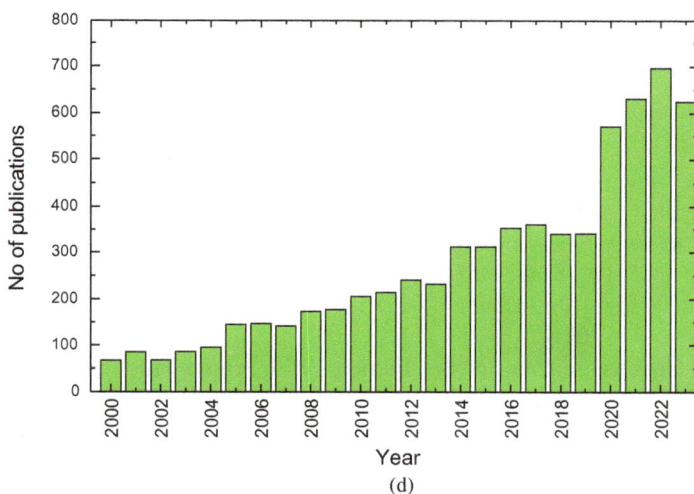

Fig. 6. (a) Publications from all Institutions of HBNI, (b) publications from BARC, (c) publications from IPR, and (d) publications from TMC.

Epilogue

The decision to start the Training School and provide academic recognition to the training programme under the umbrella of HBNI has been very fruitful. Graduates of the Training School have contributed to the programmes of atomic energy and "several other major scientific and technological efforts outside the ambit of atomic energy" [9].

In recent decades, the size and cost of experimental facilities are increasing, and constructing, running and maintaining facilities require well-trained manpower, which is available in large national laboratories. This kind of manpower is not available in Higher Educational Institutes in India and also in most other countries. One way to address this gap is to cluster national laboratories and universities as has been done in the case of HBNI where functions of a university and national laboratory are combined in the same entity. This has also been done in France by establishing several clusters such as the Paris-Saclay University.

Publications from units of DAE have been continuously increasing. After the setting up of HBNI, the rate of increase has seen an upswing as seen from the data in Fig. 6.

Contributions by Raja Ramanna in nurturing the Training School in its formative years and supporting the proposal of establishing HBNI are noteworthy.

References

1. H. J. Bhabha, General plan for atomic energy development in India. *Conference on Development of Atomic Energy for Peaceful Purposes in India*, New Delhi, November 1954.
2. H. J. Bhabha, The role of atomic power in India and its immediate possibilities. *In International Conference on the Peaceful Uses of Atomic Energy*, Geneva, 1, 103–109 (1955).
3. N. Mukunda, Homi Bhabha – an appreciation and a hope, *Curr. Sci.* 100(9), 1437–1441 (2011).
4. H. J. Bhabha, Science and the problems of development, A Lecture delivered at ICSU Conference, Bombay, 1966.
5. BARC Training School, Mumbai, Diamond Jubilee 1957–2017, accessed at https://www.barc.gov.in/publications/tshm_dj.pdf on 14 Oct 2024.
6. R. B. Grover *et al.*, From the training school to the Homi Bhabha National Institute, *Curr. Sci.* 123(3), 441–450 (2022).
7. R. B. Grover, Opening up of international civil nuclear cooperation with India and related developments, *Progress in Nuclear Energy*, 101, 160–167 (2017).
8. R. B. Grover, The idea of a university, *Curr. Sci.* 127(7), 75–76 (2024).
9. B. A. Dasannacharya, Homi Jehangir Bhabha. *Curr. Sci.* 96(11), 1536–1538 (2009).

Dr. Ravi Bhushan Grover is a Member of the Atomic Energy Commission and Chairman of the Board of Research in Nuclear Sciences. He graduated from the Delhi College of Engineering, Delhi University, and received a Ph.D. from the Indian Institute of Science. He was the President of the Indian Society of Heat and Mass Transfer during 2010–2013. He worked in the Bhabha Atomic Research Center (BARC), the Secretariat of the Department of Atomic Energy (DAE) and the Homi Bhabha National Institute (HBNI). Positions held by him include Director, Knowledge Management Group, BARC; Director, Strategic Planning Group, DAE; Principal Advisor, DAE; Director/Vice-Chancellor, HBNI; and DAE Homi Bhabha Chair.

Working in BARC, he specialized in nuclear reactor thermal hydraulics, process design and safety analysis. He participated in negotiations with other countries and international agencies leading to the resumption of international civil nuclear trade.

He played a very significant role in conceptualizing and establishing HBNI and concurrent with other responsibilities, he was its Founder Director/Vice-Chancellor from 2005 to 2016.

He was conferred with a Padma Shri in 2014. He is a fellow of the Indian National Academy of Engineering, the Maharashtra Academy of Sciences and the World Academy of Art and Science.

https://doi.org/10.1142/9789819814435_0007

Chapter 7

Dr. Raja Ramanna: An Exceptional Physicist, Able Administrator and Humanitarian

Dilip D. Bhawalkar

Raja Ramanna Center for Advanced Technology, Indore, India

ddbhawalkar@yahoo.com

I vividly recall my first meeting with Dr. Raja Ramanna in the summer of 1967. At that time, I was a lecturer in the Electronics Department at the University of Southampton, UK, teaching a postgraduate course on lasers. Earlier, I had responded to a circular from the Department of Atomic Energy (DAE) sent to Indian faculty in electronics departments at UK universities, inviting applications from those interested in joining the DAE. I had nearly forgotten about my application when, to my surprise, I received a telegram asking me to meet Dr. Ramanna at the Strand Hotel in London in May 1967 (the exact date escapes me). On the appointed day, I arrived at the hotel and met Dr. Ramanna, who was then the Director of the Physics Group at the Bhabha Atomic Research Centre (BARC), Trombay, Mumbai. He informed me that he only had 30 minutes and asked me to explain my work. Lasers were a relatively new field at the time, so I did not expect him to be very familiar with it. To my astonishment, as I explained to him my research, he quickly grasped the subject and posed insightful, pointed questions. Our conversation extended well beyond the originally planned 30 minutes. The most significant outcome of that meeting for me was that I joined BARC in December 1967.

While at the University of Southampton, I had come across a report from the Indian High Commission in London detailing a laser communication link that BARC scientists had established between Trombay and Tata Institute of Fundamental Research (TIFR), Colaba, Mumbai, spanning more than 20 kilometres. I was deeply impressed by this work and was eager to join the Electronics Division at BARC to contribute to such projects. By the time I joined, however, Dr. M. B. Khambatti, who had led this initiative, had left, and I took over the leadership of the project. Initially, I did not interact much with Dr. Ramanna since I was in the Electronics Division of BARC. But that changed when Dr. Ramanna became Director of BARC in 1972. I began to meet him more frequently, particularly as he showed keen interest in the development of a pulsed 10 GW peak power Nd Glass laser designed for generating and studying high-temperature, high-density plasmas. After seeing the laser in action, he called me to his office to discuss the possibility of using high-power lasers to excite uranium nuclei. I explained that typical photons have energies of about 1 eV, while exciting a uranium nucleus would require photons with energies in the MeV range. Although multiphoton absorption was theoretically possible, the probability of achieving excitation through millions of photons was infinitesimally low. He accepted this reasoning. But his idea, though premature then, is coming true today. Interestingly, scientists today use Compton scattering of high-power infrared lasers on high-energy electron beams to produce variable-energy photons in the MeV range, enabling them to measure the cross-sections for uranium nucleus excitation.

One of Dr. Ramanna's most admirable initiatives was his practice of having a meeting regularly with younger scientists to discuss their ideas. We all appreciated and enjoyed these stimulating and informal discussions.

As Director of BARC, Dr. Ramanna recognised that the facility had limited space to accommodate new activities requiring large areas. In the early 1980s, he decided to establish a new research centre dedicated to R&D in fields that were not directly related to the nuclear program. Initially, the centre was intended to focus on particle accelerators, lasers and plasma physics. Dr. Ramanna took a deep personal interest in the project and worked diligently to secure approval for the proposal from the Prime Minister.

While approving the proposal, Prime Minister Mrs. Indira Gandhi added a crucial stipulation: the new institution should be located in one of

India's scientifically less-developed states. To find a suitable site, Dr. Ramanna formed a site selection committee, which he chaired and he himself visited several potential locations in Uttar Pradesh, Madhya Pradesh, and possibly Bihar. After a thorough evaluation, the committee identified Indore, where the Madhya Pradesh government had offered nearly 800 acres of land. Dr. Ramanna then called me to his office. Since I am from Sagar in Madhya Pradesh, he asked me to visit the proposed site and offer my suggestions. A few days later, I travelled to Indore to inspect the land. I found that the site was fragmented, with a major road cutting through it. Moreover, several small colonies, temples and mosques occupied part of the land, and I felt that some of these structures might be difficult to relocate. As a nature enthusiast, I had hoped for a site with more natural beauty. I requested that the local authorities show me alternative locations, and they took me to nearby hills. They mentioned that this site had already been considered by the committee and rejected – an outcome I would have agreed with (Fig. 1).

Fig. 1.   Dr. Raja Ramanna (with binoculars) and the author (second to his left) at the site of future RRCAT.

However, as we continued further, I noticed a water body. Curious, I inquired about it, and the authorities explained that it was Sukhaniwas Lake, with a palace on its shores. I asked whether this area could be made available for the centre. At 11:00 AM, a message was sent to the Mayor of Indore, who called an emergency meeting of the city corporation. By 3:00 PM, I received confirmation that the lake, the palace and the surrounding land could indeed be allocated for the project. I returned to Mumbai and briefed Dr. Ramanna on this new development. The committee reviewed the site and gave its approval.

On 19 February 1984, President Giani Zail Singh formally inaugurated the centre, which was named the Center for Advanced Technology. I am proud that today, this institution has been renamed the Dr. Ramanna Center for Advanced Technology, a fitting tribute to a remarkable physicist and visionary who played such a pivotal role in establishing this research centre (Fig. 2).

Fig. 2.   President Gyani Zail Singh (fourth from left), Dr. Raja Ramanna (extreme left) and C. Ambasankaran (Chairman of Planning and Implementation Committee, extreme right) during the inauguration of Centre for Advanced Technology, later renamed as Raja Ramanna Centre for Advanced Technology.

Beyond being an exceptional scientist, Dr. Ramanna was a compassionate leader who genuinely cared about the well-being of those around him. In 1981, I began to lose my vision and was diagnosed with a pituitary adenoma that was pressing on my optic nerve, threatening further loss of vision. My neurologist recommended immediate surgery in the USA.

Although Dr. Ramanna was then the Secretary and Director General of DRDO, he reached out upon hearing of my condition and arranged for Air India to provide tickets for my wife and me to travel to the USA for the procedure. Dr. Ramanna's kindness and support during that challenging time is something I will always cherish.

Dr. Dilip Devidas Bhawalkar is an Indian Optical Physicist and the Founder Director of the Centre for Advanced Technology (CAT) at Indore later renamed as Raja Ramanna Centre for Advanced Technology (RRCAT). Dilip Bhawalkar obtained his B.Sc. (1959) and M.Sc. (Physics) (1961) from Sagar University, after which he earned a master's in Electronics (1963) and Ph.D. in Laser Physics (1966) from Southampton University. He joined Southampton as a Lecturer but returned to India to join the Bhabha Atomic Research Centre (BARC) in 1967. Dr. Bhawalkar is one of the pioneers of laser in India and one of the early doctoral scholars in the technology when the discipline was at its nascent stage. During this period, he was instrumental in setting up various laboratories and facilities for CAT and developing a National Laser Programme for the country. He is credited with pioneering research in optics and lasers in India. During his tenure, CAT became a partner in the International Linear Collider and Large Hadron Collider experiments of the European Organization for Nuclear Research (CERN).

Dr. Bhawalkar received the Shanti Swarup Bhatnagar Award (1984), University Grants Commission National Lecturer Award (1984), Goyal Award for Physics (1997) and Firodia Award (2000). He was elected a Fellow of Indian Academy of Science, Bengaluru; National Academy of Sciences (India), Prayagraj; Optical Society of America; and a foreign fellow of Russian Academy for Natural Sciences. The Government of India awarded him the fourth highest civilian award Padma Shri in the year 2000.

Chapter 8

Construction of VEC: Beginning of the Era of Large Accelerators in India

Rakesh K. Bhandari

Inter University Accelerator Centre, New Delhi, India

Variable Energy Cyclotron Centre/DAE, Kolkata, India

rakeshbhandari808@gmail.com

Abstract: Construction of the Variable Energy Cyclotron (VEC), indigenously, during the 1970s was a challenging task considering the industrial and technological infrastructure available in the country. It is an azimuthally varying field (AVF) cyclotron – a state-of-the-art accelerator capable of delivering a variety of particle beams over large energy ranges. To build this machine was a foresighted decision of Raja Ramanna and Homi Bhabha in the early 1960s as they knew that the cyclotron is a versatile machine, and it would allow the nuclear physics community to carry out very good quality research to keep pace with the international community. Other areas of research such as radiation chemistry, radiation physics and radiation biology also would benefit. This chapter is an inspiring story of how Indian scientists and engineers, in collaboration with the industry, accomplished the difficult task of constructing a large yet sophisticated accelerator, over 5 decades ago, in less than 7 years' time. A premier centre for advanced research in accelerator-based nuclear sciences was made a reality, in a then remote location of Calcutta, fulfilling the dream of Ramanna and Bhabha.

Introduction

India's top nuclear physicists got together in a meeting at IISc, Bangalore, in August 1964 to deliberate on the next particle accelerator in the country for experimental research. Until then, the biggest accelerator available was a 5.5 MV Van de Graaff machine installed in 1962 at the Bhabha Atomic Research Centre (BARC), Trombay (Fig. 1). Homi Bhabha summed up the decisions of the meeting as follows [1]:

> *I think the result of this meeting has been extremely instructive and very fruitful. I am convinced that there is a large field of operation in nuclear physics where fruitful work can be done with these machines. The following general situation seems to emerge in my mind. One is that the two machines, the Tandem and the AVF accelerator are complimentary. It may be possible for an AVF machine with the help of elaborate equipment to do what a Tandem does. If we decide to go in for a Tandem, it seems clear that we should buy one. The technology of an AVF is entirely different and this is the field, except for the cyclotron at Calcutta, which has been neglected in India. This is the field which I think should now be developed. We are in a position now to enter a new field of building accelerators. Therefore, I myself favour the idea of going in for an AVF machine.*

Discussions at the IISc meeting centred around the recommendation of the Raja Ramanna committee that was constituted in 1962. It had recommended a Tandem machine and/or an AVF cyclotron for the Indian nuclear physicists. In Bhabha's mind, the merits of building an AVF cyclotron were, apparently, the development of useful technologies such as large electromagnets, high-power radiofrequency (RF) systems and large vacuum systems in the country.

Eventually, the Atomic Energy Commission (AEC) approved the construction of a large AVF cyclotron, indigenously, in 1969. The project was to be executed at Calcutta by the scientists and engineers of Bhabha Atomic Research Centre. A deserted patch of land measuring about 22 acres (Fig. 2) was made available by the West Bengal government in a corner of a new township called Salt Lake City (now Bidhan Nagar). This land would house the "Cyclotron Centre of BARC" as well as the new campus of Saha Institute of Nuclear Physics (SINP). The visitors to the site used to be welcomed by swarms of mosquitoes, snakes and other wild animals. Accessibility also was a big issue. Execution of the

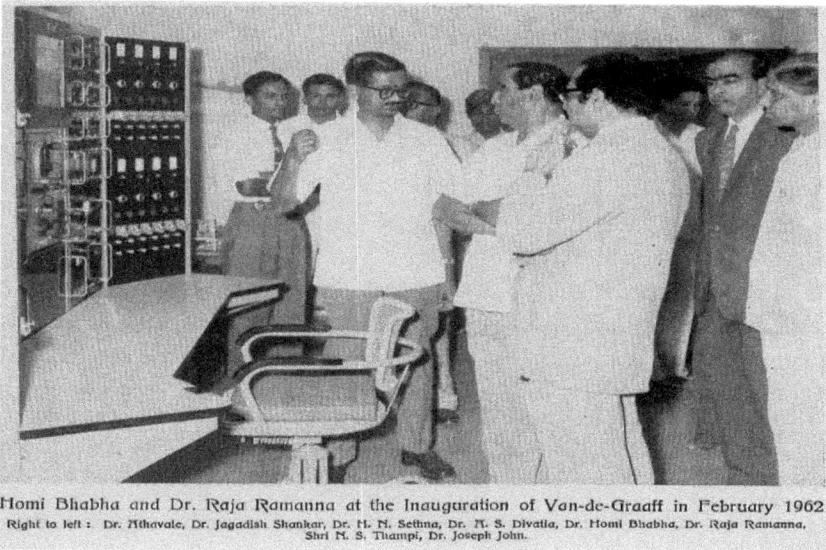

Homi Bhabha and Dr. Raja Ramanna at the Inauguration of Van-de-Graaff in February 1962
Right to left : Dr. Athavale, Dr. Jagadish Shankar, Dr. H. N. Sethna, Dr. N. S. Divatia, Dr. Homi Bhabha, Dr. Raja Ramanna, Shri M. S. Thampi, Dr. Joseph John.

Fig. 1. Van de Graaff control room at BARC (February 1962).

Cyclotron Building under construction at Bidhan Nagar, Calcutta

Fig. 2. VEC building under construction (~1971 photo).

VEC project was a challenging task in many ways as we shall see in the following sections.

Variable Energy Cyclotron (VEC)

As mentioned, VEC is an azimuthally varying field (AVF) cyclotron that can accelerate a wide variety of ions over large energy ranges. The maximum energy achievable is K.(Q^2/A) MeV, where Q and A are the charge and mass numbers, respectively, of the ion. Factor K is a measure of the maximum bending power (BR) of the cyclotron magnet where B is the maximum magnetic field in the accelerating gap and R is the radius of the magnet pole. In the case of VEC, the K value is 130. Since it is a room-temperature magnet, the maximum magnetic field is ~1.7 T and R is about 1 m. Thus, VEC is a very large cyclotron. In fact, its pole diameter is 2.24 m. The total weight of the magnet is ~260 tons including poles and return yoke. Azimuthally varying field is achieved by attaching large spiral sectors of the poles – 3 on the upper and 3 on the lower pole (Fig. 3). Machining and assembly tolerances are in microns – making fabrication extremely difficult. The other complex system is the radiofrequency (RF) system which must deliver power in the range of ~400 kW to generate enough accelerating voltage for the ions. The frequency range is 5.5–16.5 MHz. This decides the energy range over which the ions

Fig. 3.   Lower part of the VEC magnet frame with spiral sectors attached to the magnet pole (~1974 photo).

can be accelerated. For example, the energy range for deuterons is 12–65 MeV and for alpha particles 25–130 MeV. The cyclotron vacuum chamber has a volume of about 33 m^3 in which a vacuum level of better than 10^{-6} mm of Hg needs to be maintained.

Design drawings for the cyclotron were obtained from Texas A&M University, Texas, while the original design was done by the Lawrence Radiation Laboratory, Berkeley. However, planning and design of the beamlines were done by VEC experts. All these cyclotrons, VEC included, continue to operate with their own add-ons and upgrades to date for experiments.

Execution of the VEC Project

Right from the conceptual stage in the early 1960s to realisation in 1977, Ramanna was playing a leading and inspiring role in the cyclotron project. In fact, the selection of the machine to be built for India was done under his careful supervision. After the project was approved in 1969, he took over as Chairman of the Variable Energy Cyclotron Committee, the VEC Committee, the highest policy-making body (Fig. 4). It consisted of top scientists of the country. He was Director of the Physics Group of BARC until 1972 and then Director of BARC until 1978.

Fig. 4. VEC Project Coordination Committee meeting in progress with Raja Ramanna (right row, fourth from right) in chair at the site (~1973 photo).

For the actual execution of the project, a Project Coordination Committee (PCC) was constituted under the chairmanship of C. Ambasankaran who was Director of the Electronics and Instrumentation Group of BARC. Other members of PCC were senior scientists and engineers from BARC with different specialisations. D. Y. Phadke, who was a senior member of TIFR, was associated with the project as a highly knowledgeable mentor.

A. S. Divatia, who was a senior nuclear physicist with accelerator technology background at BARC, played the most important role in making VEC a reality. He led the project as Project Manager and later Project Director while remaining stationed at Calcutta for almost the entire duration of the project. He accomplished the tough job with his sound expertise, international connections and personal charm. He later took over as Director of the VEC Centre (VECC).

Difficulties, Challenges, Innovations

Accelerators require high precision in fabrication and assembly. Constructing VEC indigenously was particularly challenging because the technical infrastructure in the country was still developing at that time. Several technical solutions had to be evolved to build the cyclotron systems with acceptable specifications and performance as described in the following sections. Furthermore, the VEC team members had to learn cyclotron physics and technology, virtually, from scratch. Of course, there was some technical help available from the Texas and Berkeley cyclotron institutes. Some of the team members were deputed to those institutes for short durations. This exposure certainly helped.

Magnet Frame

Magnet frame, a high precision component, is made of 12 large steel pieces each weighing ~20 tonnes. Very high-quality steel was required with low carbon contents and minimal blow holes. Fabrication of the frame using forged steel was recommended. India, at that time, did not have facilities to forge such large pieces of steel. Imports were to be avoided or minimised. The matter was discussed with the experts at M/s. Heavy Engineering Corporation (HEC) at Ranchi. It was decided

Fig. 5.   Dr. S. Chatterjee with the magnet iron frame along with the main coils assembled at site (~1975 photo).

that each of these pieces would be cast with a large amount of extra steel – about 50 tonnes [2]. The extra steel in the riser portion on the top of the cast piece would be machined off to get rid of the majority of blow holes. This technique worked well, and excellent quality steel was produced with acceptable blowhole contents and very good magnetic properties. Forged steel for the smaller components such as pole pieces (~10 tonnes each) and spiral sectors was imported. The machining of all the pieces of the magnet frame was done in various machine shops of M/s. HEC with unprecedented quality to meet the strict assembly and surface finish toler-ances. This was not so straightforward with the vendor whose preference was to churn out steel pieces and structures rather than high accuracy. Machining of the spiral sectors and their assembly with the rest of the frame involved sophisticated techniques. The precision of the entire job was a challenge for M/s. HEC that they successfully met with the involve-ment of VEC engineers. The total weight of the magnet frame assembly is about 260 tonnes (Fig. 5). It was a marvellous feat in the early 1970s in the country.

Magnet Coils

There are three different types of coils for the electromagnet. These are the main coils, trim coils and harmonic correction coils. Only M/s. Bharat Heavy Electricals Limited (BHEL), Bhopal, had the infrastructure to fabricate these coils. A major problem arose in fabricating the concentric trim coils – 17 coils on the face of the upper pole and 17 on the lower pole. They carry excitation currents ranging from 750 to 2500 A consuming a total of up to ~500 kW power. They were to be made using mineral-insulated hollow copper conductors. This special conductor was not available in India. Moreover, it was on the list of banned items to be exported to India. At the same time, the coils are crucial for trimming the radial profile of the magnetic field for isochronism or else the beam would be out of phase with respect to the sinusoidal accelerating field and lost. Once again, the experts of the VEC project and M/s. BHEL came up with an excellent solution of epoxy potting the coil that finally worked. Considering the size of the assembly (Fig. 6), it sure was a very difficult job. Epoxy had to be thoroughly tested for radiation resistance properties by exposing samples to high neutron flux in the APSARA reactor at BARC. M/s. BHEL also, successfully, fabricated the main coils that used a 28 mm × 28 mm cross-section copper conductor with a 19 mm diameter

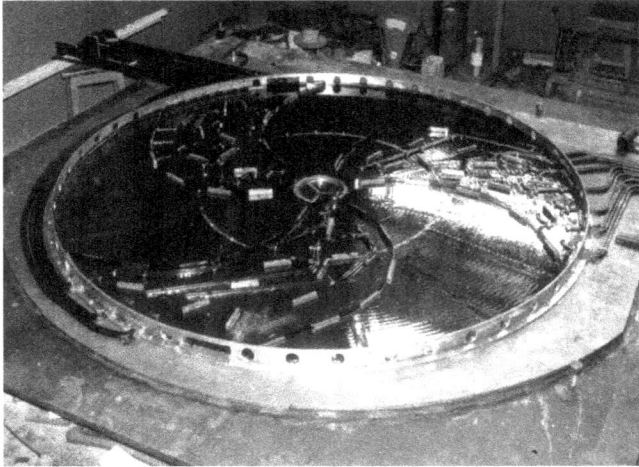

Fig. 6.   Assembly of epoxy-potted trim coils and valley coils of VEC under fabrication at BHEL (~1975 photo).

central (Fig. 5) hole for cooling water flow. Up to ~1 MW of power is dissipated in the main coils.

It was not an easy job to fabricate power supplies for all these coils. Main coils are excited by a power supply delivering up to 2800 A current with long-term stability of better than 10^{-4}. Stability requirements are similar for the trim coil supplies.

Radiofrequency (RF) System

This 400-kW system was no less challenging to make at that time in the country. Huge copper resonators and dee-dee stem assembly (Fig. 7) required fabrication using furnace brazing to avoid hot spots during operation. No such facility was available in India at that time. The Central Workshop of BARC completed the job by skillfully using conventional brazing techniques. These resonators are housed in a small room-sized chamber (Fig. 8) evacuated to less than 10^{-6} mm Hg pressure. This chamber was fabricated by a shipbuilding company, M/s. Garden Reach Workshops, Calcutta, under the strict supervision of the vacuum technology experts of BARC and VEC projects. Building the 400-kW electrical

Fig. 7. Dee-dee stem assembly being prepared at VEC for operation. Raja Ramanna can be seen in the picture (~1976 photo).

Fig. 8.   Resonator tank of VEC being lowered in position (~1975 photo).

part of the RF system along with complex power supplies was no less a challenge that was successfully met by the VEC team. It is a vacuum tube-based system, and the tube had to be imported while all the power supplies were built in-house.

Vacuum System

Achieving very low pressure, $<10^{-6}$ mm of Hg, in the cyclotron for proper operation also posed a difficult challenge. As mentioned, the total volume to be pumped is about 33 m^3. Oil diffusion pumps were the best solution. Design calculations showed that two 90 cm pumps would be required. India was probably not making bigger than 15 cm pumps at that time. BARC experts took up the challenge and designed and fabricated the

Fig. 9. 90 cm oil diffusion pump for VEC. It was equipped with freon-cooled baffle at the top (~1976 photo).

required pumps (Fig. 9) along with huge gate valves. Back streaming of oil into the cyclotron chamber with such huge pumps also was a big issue that was resolved.

Physics of the Cyclotron

Beam dynamics of AVF cyclotrons is highly complex particularly when the sectors are of spiral shape. Furthermore, highly accurate (<1 part in 10^{-4}) magnetic field measurements, amounting to over one million data points, are required to generate operational settings. Relevant computer codes and techniques were developed by our physicists and engineers. The job was particularly challenging as it was to be done, virtually, from scratch using rather primitive computational facilities available in the country in the early 1970s. Some help was forthcoming from the Berkeley and Texas colleagues. Using the above knowledge, the first beam was accelerated in VEC, after short trials, on 16 June 1977 (Fig. 10).

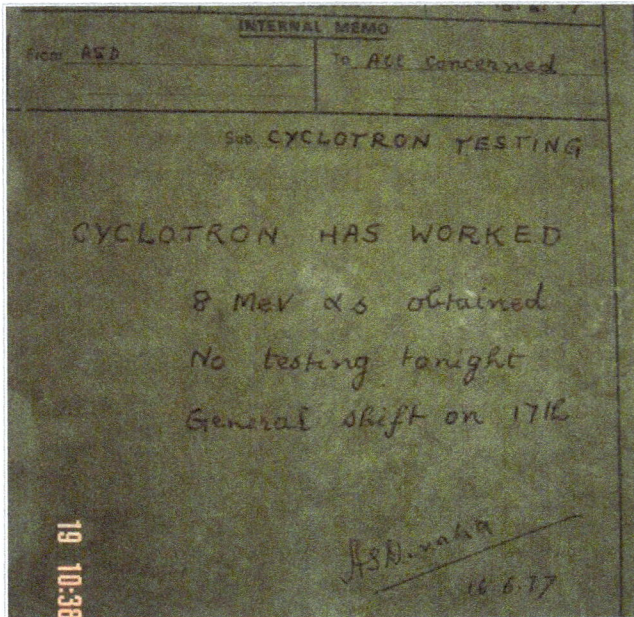

Fig. 10.   VEC accelerates first beam.

Electrical Power Issues

The story of VEC will not be complete unless mention is made of the electrical power situation in Calcutta when the VEC project was being implemented. The supply was irregular, and the quality was unacceptable for the cyclotron systems. The situation was a bit better during the night-time. The teams used to work during those hours for systems' tests and commissioning. Even commissioning of the cyclotron, with all systems operating in unison, was done with the teams working at nighttime for months at a stretch. Four 500 MVA diesel generators were later installed to run the machine. Immediately after that, the problem was permanently resolved when dedicated power lines for VEC were installed. This was due to the persistent efforts of Ramanna with the local government.

Other Issues

VEC was a greenfield project with a dedicated team for its execution. Despite the limitations of technical infrastructure in the country and various other problems as mentioned above, the cyclotron was commissioned

Fig. 11. VEC with its beamlines.

in less than 7 years' time. Majority of the manpower had to be moved from Bombay to Calcutta. Furthermore, in those days, letters, telexes and telegrams were the primary modes for communicating information. The cyclotron (Fig. 11) continues to operate to date, delivering light and heavy ion beams, for experiments. In the subsequent years after 1977, several improvements and upgrades and additions were made to the VEC.

Experimental Facilities and Physics Research

Raja Ramanna inspired the community of nuclear physicists and other prospective users of VEC to plan, well in advance, the research facilities that would be developed and installed at VEC for experiments. He wanted high-quality experiments to begin as soon as the VEC delivered the beam. Like the cyclotron systems, the emphasis would be on developing the research facilities, indigenously. A national-level group comprising scientists from universities and national research centres was formed to accomplish this task. This was, essentially, an advisory group and implementation of their recommendations was to be done by the VEC teams. Dedicated manpower that consisted of physicists and engineers was assigned for this purpose in the project.

Internationally acclaimed nuclear physicist N. K. Ganguly took over the leadership to implement this activity for the VEC. Advanced laboratories to develop precision nuclear electronic modules, data acquisition systems, computational facilities, solid state detectors and nuclear targets were set up apart from developing data acquisition codes for the experiments. After he laid down the office, another renowned physicist, Bikash Sinha, carried forward this activity and became instrumental in setting up still more advanced accelerators and research facilities at the VEC Centre for Nuclear Physics Research. He, subsequently, became Director of the Centre in 1984.

Another important aspect of the VEC project was to set up strong physics groups, both experimental as well as theoretical, in-house for a seamless transition from the project phase to the research phase. Physics research on cyclotron energies was relatively new to Indian scientists. Very young physicists such as Dr. D. K. Srivastava and Dr. Y. P. Viyogi, who had joined N. K. Ganguly's team in the early stages of the project, went on to become scientists of international fame.

Era of Large Accelerator Facilities Begins In India

Though, with the construction of VEC, Ramanna's dream of significantly raising the platform of nuclear physics research in India had been fulfilled, he was still thinking ahead. During his frequent visits to Calcutta, he used to interact with the accelerator and research scientists and inspire them to plan for the next higher steps. Success of VEC had generated enough confidence to take up challenges of building still bigger accelerators in the country. Furthermore, the scientific communities that wanted to do accelerator-based experiments in the country were growing fast. The nuclear physics community was mostly interested in high-energy heavy ion beams. Scientists engaged in other areas of research such as high energy physics, material sciences, condensed matter physics and biology also looked for advanced accelerator facilities for experiments. In March 1979, a national-level committee was set up under the chairmanship of P. K. Iyengar, Director, Physics Group, BARC by H. N. Sethna who was at that time the Chairman, AEC & Secretary, DAE, Government of India. It was a very high-level committee with members from DAE, SINP, academic institutions, Planning Commission and Department of

Science & Technology. M. K. Mehta, a renowned nuclear physicist, was the Member-Secretary. The committee held several meetings and organised workshops for the prospective user groups. Working groups on different types of accelerator facilities *viz.* accelerator for nuclear physics with heavy ions, GeV accelerator for high energy physics and synchrotron radiation sources were constituted. The entire exercise went on for over 2 years. Finally, it was decided to set up a large synchrotron radiation source facility comprising multiple sources. Ramanna convinced the Madhya Pradesh government to provide a huge stretch of land for this greenfield facility. Thus, the Centre for Advanced Technology (later renamed Raja Ramanna Centre for Advanced Technology) was created. It was decided that this Centre would also have, apart from accelerators, large-scale activities to develop advanced laser systems.

In the subsequent years, accelerator facilities continued to develop at many centres in the country. VECC built a large superconducting cyclotron for GeV heavy ion beams. TIFR/BARC and Nuclear Science Centre (now Inter University Accelerator Centre) developed superconducting linacs as boosters for their pelletron accelerators for 100s of MeV heavy ion beams. Extensive R&D is underway on accelerator-based spallation neutron source and Radioactive Ion Beam (RIB) facilities.

Summary

Indigenous construction of variable energy cyclotron in the first half of the 1970s was a milestone step in the history of particle accelerators in India. This dream of Homi Bhabha and Raja Ramanna was successfully realised by Indian scientists and engineers. Advanced nuclear physics research was the motivating factor. Inspired by Ramanna, several other large accelerator projects were implemented in the country in the next several decades.

Acknowledgement

The author is grateful to Prof. V. S. Ramamurthy and Prof. D. K. Srivastava of NIAS, Bengaluru, for their encouragement to write this short story of Variable Energy Cyclotron (VEC).

References

1. Divatia, A. S. and Ambasankaran, C., Accelerator Development in India. *Pramana*, 24(1 & 2), 1985, 227–224.
2. Ambasankaran, C. and Phadke, D. Y., 224 cm Variable Energy Cyclotron at Calcutta. *IEEE Transactions on Nuclear Science*, NS-20(3), 1973, 236–239.

Dr. R. K. Bhandari is a leading accelerator physicist and technologist of the country. He was associated with the construction of the K130 Variable Energy Cyclotron (VEC) – the first large accelerator built in India. Subsequently, he played a leading role in setting up of the K500 superconducting cyclotron and 30 MeV medical cyclotron facilities in Kolkata. He was Director of the Variable Energy Cyclotron Centre (VECC) until June 2012 and held the position as Distinguished Scientist. Presently, he is associated with the Inter University Accelerator Centre (IUAC), New Delhi as an Hon. Visiting Scientist.

Chapter 9

Dr. Raja Ramanna: My Reminiscences

K. S. Parthasarathy

Atomic Energy Regulatory Board, Mumbai, India

Department of Atomic Energy, Mumbai, India

ksparth@gmail.com

Preamble

I saw Dr. Raja Ramanna for the first time when he came to visit the Atomic Energy Establishment Trombay (AEET) Training School hostel within a week or so after I joined the school. A day or two earlier, Shri Crasto, our hostel warden informed us that Dr. Ramanna was visiting us, and we must assemble in the hostel hall at least 15 minutes earlier. Later, I learned from trainees of the earlier batches that Dr. Ramanna has been carrying out this practice of meeting the trainees of every batch in the hostel.

While in the Training School, I met him face-to-face on a few occasions. However, after joining the Atomic Energy Regulatory Board (AERB) in late 1984, I interacted with him more closely. When Dr. D. K. Srivastava and Dr. V. S. Ramamurthy asked me to contribute, my reminiscences about Dr. Ramanna, I accepted their proposal with mixed feelings. I retired on 31 January 2004. Since 1 April 2004, I continued to serve the Department first as a Board of Research in Nuclear Sciences (BRNS) Senior Scientist and later as a Raja Ramanna Fellow for eight years. I felt very happy when the fellowship name changed. I recalled the occasions

I interacted with Dr. Ramanna. I have explained them over the years in a few articles. These articles and other information I had collected from other sources including from the experiences of some of my senior colleagues became my resource to write this chapter.

What were Dr. Ramanna's major contributions? While paying tribute to Dr. Homi Bhabha, Mr. J. R. D. Tata observed, "I believe that the greatest contribution Homi made to India's development into the modern state it is fast becoming, lies in training and bringing out to their full capability a host of young scientists and administrators who, today, lead so many of India's scientific and technical establishments".

Among the young scientists handpicked by Dr. Bhabha, Dr. Ramanna stands out as a shining example. Dr. Bhabha assigned him the responsibility to organise a training programme at the Atomic Energy Establishment Trombay (AEET). Dr. Ramanna proposed that the school may be called the Atomic Energy Establishment Trombay Training School (AEET TS).

Dr. K. K. Damodaran, former Head of the Training Division, who assisted Dr. Ramanna in nurturing the training programme, remembered that one of the mandates of the school was to take steps to attract young, bright and talented students from the universities. During training, they acquired the needed skills and knowledge in nuclear science and technology. The need-based, dynamic curriculum ensured that the trainees would be well prepared to face the challenges when the technology gets upgraded or when new technology appears on the scene.

Why did Ramanna have a soft corner for trainees? Ramanna suffered hardships during his formative years in England.

In 1944, Homi Bhabha was impressed when he saw Ramanna playing the piano at the State Government Guest House, Mysore. Shortly thereafter, Bhabha recommended Ramanna for the J. N. Tata Endowment for the Higher Education of Indians for a scholarship to enable him to go to England for a Ph.D. in Physics.

Dr. Bhabha and Dr. Ramanna knew that since nuclear technology is a strategic technology, the free flow of knowledge and materials would not be forthcoming. They were conscious of the long-term need for self-reliance.

According to Dr. P. K. Iyengar, former Chairman of the Atomic Energy Commission (and long-term associate of Dr. Ramanna), Dr. Ramanna used to say that the selection of trainees is essentially a "statistical operation". Dr. Ramanna believed that in a developing country like India, if we want to get the most talented people, we must choose every year a few hundred from the vast pool of academically sound young people. If ten out of two hundred turn out to be outstanding, the selection process will be successful.

The systematic recruitment of outstanding young people year after year is the reason for the success of the BARC Training School programme. A significant number of those selected became leaders in science and technology as they enhanced their analytic skills and creativity in the multidisciplinary ambience provided at Trombay. Dr. Ramanna's characteristic humility forbade him from waxing eloquently on this notably successful human resource development programme. In his autograph, *Years of Pilgrimage*, Dr. Ramanna spared eight sentences to describe some of its features!

First Batch Training School Course

In August 1957, Dr. Bhabha and Dr. Ramanna recruited 143 trainees: 49 in the engineering stream and 104 in the science stream.

Dr. P. S. Nagarajan, who belonged to the first batch, vividly remembers their first encounter with Dr. Ramanna. The Administrative Officer asked the trainees to assemble near the dining hall within two days of their arrival. Dr. Ramanna was scheduled to address them.

Nobody noticed when a young officer came, stood near the small table nearly reclining against it and started talking. Nothing was audible as there was too much noise. One of the trainees approached the officer and told him that they were waiting for Dr. Ramanna. "I am Dr. Ramanna", the young speaker revealed. Nobody could believe that the person who looked like a college boy was indeed Dr. Ramanna.

Dr. Ramanna chaired the Training School coordination committee from the very beginning. Dr. K. K. Damodaran was its Member-Secretary. Drs. A. S. Rao and Jagdish Shankar and other eminent scientists were members of the Committee. The Member-Secretary enjoyed the powers of the Head of a Division to carry out the daily administration of the school.

Dr. G. Venkataraman, also belonging to the first batch, remembers that he was made Mess Secretary by Dr. Bhabha and so Dr. Ramanna knew his name. Once Dr. Ramanna brought a person to our hostel and said, "Venkataraman, this is Iyengar, just returned from Canada. Tell him about the Training School". Thus, Venkataraman got a chance to talk to Dr. P. K. Iyengar.

Whenever the trainees faced any difficulty in the Training School, Dr. Ramanna gladly offered guidance and advice. He used to tell them that knowing the subject is different from understanding it. By the time the second batch of trainees joined in August 1958, a well-organised programme was in place. Thereafter, all trainees received training in Mumbai itself.

Most of us were used to one or two examinations over the year, not weekly examinations, take-home assignments, periodic viva voce, tutorials and lectures at such a rapid rate; life was too busy.

In our formative years, few had occasion to interact with Dr. Bhabha closely. Dr. Ramanna was different. He was our mentor. At the informal meetings, he listened to us carefully and spoke quietly. He spiced his talk with funny anecdotes. Each one of us felt that he was talking to us individually. His reassuring demeanour gave us confidence.

"Dr. Ramanna was truly great, he was totally devoted to science. … he gave me a free hand", Dr. Damodaran who was intimately associated with the Training School from 1957 to 1981 gratefully acknowledges. He remembers that Dr. Ramanna used to visit the Training School and the students' hostel once in two weeks.

Dr. Ramanna's abiding interest in the training programme was a source of inspiration to all. It is interesting to speculate why Dr. Ramanna ensured that the trainees received well-organised training before they formally joined the AEET as staff. He showed a greater degree of understanding and compassion to the trainees' problems than others. His autobiography is very revealing in this context.

Ramanna's Trials and Tribulations

Ramanna arrived at Kings College, London, in September 1945 after travelling for about a fortnight by Orion, a ship which carried over 5000 troops on repatriation from various war centres in South Asia. He was among the 300 or so other passengers. They had to suffer unspeakable deprivations. They had only two door-less WCs for the 300 of them and had to queue up at odd times of the day or night to relieve themselves! To top it all, the troops were unfriendly and abusive.

After reaching London, his first interview was with one Dr. F. C. Champion ("a handsome young man in his youth but looked most severe with his thick glasses and curt manners which seemed very disturbing",

Dr. Ramanna later recalled). Ramanna felt very unhappy because Champion told him that he could register only for an M.Sc., though he had been admitted for a Ph.D. degree. I recall that this was a common problem for many who went to the UK for their Ph.D. I faced this issue for a year at Leeds University.

Ramanna used the fine art of flattery to good humour Champion. He claimed that it was easy for him as he was brought up in the Mysore court! Dr. Champion sent him to Dr. Chapman on a possible problem of establishing a correlation between the cosmic ray phenomenon and ionospheric activity. Chapman, an expert in ionospheric studies, told him that he did not see any correlation. Ramanna's persuasive skills did not work.

Shortly afterwards, he met the new head of the Department of Physics, Dr. Alan Nunn May who had worked in Canada on the British Atomic Energy project. To his great relief, May told Ramanna that registering for a Ph.D. degree would not be difficult.

Dr. May initiated Ramanna into the field of experimental nuclear physics. Ramanna's delight, however, was short-lived. Within a few days, police arrested Dr. May for leaking atomic secrets to the Russians. Dr. Ramanna went back to Dr. Champion. By then he had developed enough confidence. He worked in the basement and subbasement rooms next to the King's College hospital mortuary, all the time suffering from the smell of formaldehyde. "Occasionally on the days when we felt frustrated, we toyed with the idea of disposing of our professor and supervisor through this route", he confessed!

No wonder Dr. Ramanna was compassionate and empathised with the trainees and always lent his ear to their problems.

Dr. Ramanna initiated and nurtured a human resource development programme at such a large scale. It was unique in India. Dr. Ramanna and his colleagues took innovative steps which paid rich dividends. They used the services of the large numbers of trained scientists and engineers already available in Trombay to teach a small number of bright students recruited for the Training School. Often, the faculty exceeded the number of students! This interaction benefited the students and the teachers. The latter could concentrate on the few who had already proved their worth.

The training programme helped harmonise the standards of students from different universities. Trainees in various streams had to study some subjects which might not initially be to their liking. For instance, those in the engineering stream had to study health physics. The trainees in the science stream had to study reactor theory. The truly multidisciplinary programme prepared the trainees to face the challenges in their careers.

Dr. Ramanna used to personally participate in various stages of the training programme. He kept a spreadsheet containing the complete details including the performance of the trainees before him in the final allotment interview. In a few cases, if he felt that the performance in some subject was not up to expectation, he would ask the trainee the reason for the shortfall.

Dr. Ramanna constantly reminded the trainees about their future roles in the Department. When occasion demanded, the smiling teacher transformed into a steely, taciturn and stubborn disciplinarian.

Dr. Ramanna and his Approach to Trainees

In one of the earlier batches (1958–1959), the administrative officer decided that the trainees must wear formal attire in the cafeteria and should learn Western table manners. However, wearing ties and using a fork and spoons were anathema to many.

"Handling medu vada or idli was OK. How do you eat dry chapatti with a fork?" they complained. Some trainees defied the orders. They came in their dhotis to the cafeteria.

Those who did not have dhotis wore bed sheets. For one trainee, eating food was a ritual. During rituals, you do not wear chappals or shoes! His father brought him up like that. He could obviously not go against the wishes of his father. The issue became very hot. The attendants refused to serve trainees who did not wear formal dress. Some of the impatient and hungry youngsters went into the kitchen and tried to manhandle the attendants. Others started a noisy demonstration by beating their plates with spoons.

Dr. Ramanna visited the hostel. The trainees were very upset. He identified the "troublemakers". "You are an undisciplined lot", he thundered. "Let me tell you, you are not indispensable to the Department", he continued.

Dr. N. Subramanian, who belonged to that batch, gives a graphic description of the events. One trainee got up and responded, "The Department is not indispensable to us either". Under the circumstances, that was a bold statement. Nowadays, if challenged, even a kindergarten student may react that way. But certainly not so in the 1950s and 1960s. Dr. Ramanna's intervention helped restore peace in the hostel and everyone forgot that chapter. He tactfully de-escalated the situation.

Ramanna: No Hierarchical Structure in Research

One of his associates, Dr. S. S. Kapoor, remembers that Dr. Ramanna, who developed gas detectors, during his Ph.D. course in London, excelled in devising ingenious experimental techniques. Based on such efforts and using neutrons from the Apsara reactor, they could publish several papers in prestigious journals, such as the *Physical Review*. His group had to improvise, as their competitors in the West had better equipment.

Dr. Ramanna's commitment to basic research was total. A few days before his death, he telephoned Dr. Kapoor for help in preparing a PowerPoint presentation on basic research. He planned to address some students. Dr. Kapoor sent him the material.

Dr. Ramanna was completely involved in his work. Once they wrote a paper. He was uncomfortable about some points which emerged later.

Dr. Kapoor, then a bachelor, was staying in the Kenilworth building. There was a knock at his door one afternoon. When he opened the door, he saw Dr. Ramanna ready to discuss the points that needed clarification. Though Dr. Ramanna was a very senior scientist, he never recognised any hierarchical structure in scientific research.

I belonged to the seventh batch. Many who came from villages and small towns would like to forget the first few days in Bandra where the hostel was located. Most trainees were homesick. Travelling by suburban train to reach the Express Building in Churchgate and after a few months, Harichandrai House in Marine Lines were unsettling experiences.

During the first week, we had one of the most memorable and comforting experiences. Dr. Raja Ramanna visited us. He wore his hallmark khaki pants and white shirt. He was a very simple person. We could approach him any time. Very often, he came to the hostel. As the then Director of Physics Group, the Training School was his turf; he always defended its cause.

At the informal meetings, he listened to us carefully and spoke quietly. He spiced up his talk with funny anecdotes. When he spoke, each one of us felt that he was talking to us individually. His reassuring demeanour gave us confidence.

Ramanna: The Disciplinarian

An unforgettable incident revealed an altogether different facet of his character. The smiling teacher transformed into a steely, taciturn and stubborn disciplinarian.

Two trainees from our batch had a fight. One of them was weak but intemperate; the other one was strong, arrogant and short-tempered! They fought over some trivial issues. The weak fellow was playing table tennis when the strong man entered the sports room. After waiting for some time, he requested him to give him a chance. "You don't have to play any games, you are already strong", the weak fellow told him. Needless teasing developed into fisticuffs.

The weak man got seriously hurt. Friends intervened. Both the victor and the vanquished exchanged apologies. We thought that they had settled the issue. We realised that it had become too serious when the victim needed urgent medical help. He went to the dispensary. Our doctor promptly and dutifully reported the matter to Dr. Ramanna.

We thought that there would be some sort of inquiry. Dr. Ramanna thought differently. He could not tolerate indiscipline. He did not want to apportion the blame. "Irresponsible", "They are going to be gazetted officers in a few months". "Dismiss them both", he thundered.

The victim could not go to his hometown during the vacation. He lived on a liquid diet. It took time to heal. Dr. Ramanna changed his stand only after several trainees appealed for mercy. In the end, Dr. Ramanna, the disciplinarian, showed some compassion and saved the careers of a scientist and an engineer.

Ramanna's Kindness and Compassion

Puri (name changed), a scientific assistant who just joined the Bhabha Atomic Research Centre, missed the staff bus at Dadar that day. The taxi was too expensive. He waited at the main junction, hoping to get a lift.

He frantically waved his hand when a car with the BARC nameplate passed by. The driver ignored him. But after a few yards, the vehicle stopped. A bespectacled gentleman seated behind beckoned him.

Puri got into the front seat. "BARC laboratories should have been somewhere in the city and not far away Trombay", he grumbled.

The benevolent occupant was busy reading some books. Puri, at his ebullient best, interrupted him a few times. He annoyed the driver. Puri didn't care. His co-passenger wore an amused look.

As the car approached the BARC gate, his co-passenger tapped Puri on the shoulder. "Where do you want to get down?" "Is it ok if I drop you at the next junction?" "That is ok, gentleman", Puri responded. He had his own mannerisms.

After he got down, Puri forgot the episode. At BARC, he made friends with everyone. He was over 6 feet, head held high, he thought he was a born leader and was always in the front for every activity.

Two weeks later, the staff assembled near the Central Complex building to celebrate Dr. Bhabha's birthday. Founder's Day is a stock-taking occasion in Trombay.

Puri and his boisterous friends occupied the front row. As the Chairman of the Atomic Energy Commission (AEC), stood up to speak, Puri felt that the dignitary's face was vaguely familiar and the moment he recognised it, he hastily ran back to be out of sight! What he suffered was the mother of all shocks. For, it was the same gentleman who gave him a lift two weeks earlier – Dr. Raja Ramanna!

Dr. Ramanna's Unique Role: Divisional Reviews

During the Divisional Review Programmes that Dr. Ramanna started in BARC, the senior staff members of every Division presented their work. We attended them primarily to listen to Dr. Ramanna's delightful and erudite concluding remarks. He would cut the pretentious to size, compliment the deserving and point out areas for further study.

His incisive analysis was a treat; his acidic tongue lashed at the mediocre. He hated "slide rule" engineering! He strived for originality and creativity.

During the review of the Division of Radiological Protection (DRP), we presented papers and arranged an exhibition of the instruments we developed. We proudly displayed our publications, the way artisans flaunt their creations. About 30 reputed journals such as Mutation Research, Radiation Research, Physical Review and Health Physics published our papers.

One of our projections was on the way we used the Central Library resources such as reprints, microfilms and miscellaneous reports. Apparently, he did not appreciate the relevance of that information. We defended our decision because Dr. V. A. Kamath, head of Central Library appreciated the DRP staff's prolific use of the resources! That is a weak defence he said smilingly. Probably he was hinting at a conflict of interest in the matter!

Overall, Dr. Ramanna amply appreciated our efforts. The display included popular science articles on radiation safety-related topics in different languages. Dr. Ramanna read them calmly. To our chagrin, he found out that we pasted the article in Kannada upside down!

In 1982, DAE held an exhibition on "Radiation and Environment" at the Nehru Planetarium, Mumbai. The DRP played a major role in it. While the preparations were in full swing, Dr. Ramanna came unannounced. He wanted to make the first scientific event at the Centre a total success.

I nervously accompanied him – explaining exhibit after exhibit. There was a cartoon describing risks in everyday life. Dr. Ramanna looked at it closely. Our artist had imitated a cartoon by R. K. Laxman, I confided. I thought I was helpful. But Laxman does it better, he retorted. He had that hallmark mischievous glint in his eyes.[1]

Incidentally, Dr. Arun Kumar De, then Director of the Indian Institute of Technology, Bombay (IITB), inaugurated the exhibition. That was the first time I saw Dr. De who later became the first Chairman of AERB.

Personal Reflections of Dr. Raja Ramanna on IAEA

Positive Aspects of the Work of the International Atomic Energy Agency

In September 1997, the IAEA published a set of personal recollections which carry "a variety of views from twenty-five people who have played major roles in shaping the policies of the IAEA or have made notable contributions to its work at different periods of its history. They provide individual insights – often from a rarely available insider's perspective – into particular aspects of the development of an international organization…", Dr. Hans Blix, the then Director General wrote in the Preface of this unique book.

Dr. Ramanna was Chairman of the Norwegian Zero-Power Reactor Assembly (NORA) Project of the IAEA connected with heavy water reactors; Chairman of the Scientific Advisory Committee to the Director General, IAEA; and President of the General Conference of the IAEA in 1986.

Dr. Raja Ramanna's personal reflections vividly showed his interpretation of the developments in IAEA over 50 years. I will not be able to do justice to it if I summarise/paraphrase it. The piece truly reflects his

---

[1] A Tribute to an Outstanding Scientist PTI Feature, *The Daily Excelsior*, 28 October 2004.

transparent, no-nonsense attitude with an abundant sense of humour to various events which transformed IAEA over the past five decades.[2]

Setting up of AERB

On 15 November 1983, DAE issued a gazette notification signed by Dr. Raja Ramanna setting up the Atomic Energy Regulatory Board (AERB). He nurtured the Board in its early years. A DAE-appointed committee had recommended that step in February 1981.

Dr. Ramanna appointed Prof. Arun Kumar De as the first Chairman of AERB. Initially, Dr. Ramanna preferred AERB to give priority to enforcing radiation protection provisions among medical and industrial users of radiation as DAE SRC is already taking care of DAE installations.

AERB organised a national seminar on 14 and 15 March 1986, on "Radiation Exposures in Medical X-ray Practices: Consequences and Control" (Fig. 1). The invitees included Health Secretaries, Directors of Health Services and Directors of Medical Education from State Governments, representatives from regulatory agencies, standards organisations and professional associations, eminent radiologists, X-ray equipment manufacturers, physicians, radiological safety officers, medical physicists and other professionals.

Prof. De, the then Chairman, asked me to meet Dr. Ramanna and hand over to him a brief report on the status of medical X-ray safety in the country. I did not expect a long discussion. I was slightly nervous. To put me at ease, he spoke jocularly about what happened in his labs, many years ago. He was referring to the spread of radioactive contamination from a partly damaged source and the effort to decontaminate the labs. I have heard him talking about it on two or three other occasions. The decontamination of the labs was one of the first jobs Shri S. D. Soman who later became the Chairman of AERB did after joining the Health Physics section from TIFR where he was working in the beta spectroscopy lab.

Ramanna asked me a few questions about the activities of the Division of Radiation Protection (DRP) in the subject area. I explained the need for more field surveys. I described some inexpensive and simple ways to enhance protection in some instances. One of them is the addition

---

[2] https://www.pub.iaea.org/MTCD/Publications/PDF/Pub1033_web.pdf.

Fig. 1. Dr. Ramanna (on the lectern) inaugurating National Seminar on Radiation Exposures in Medical X-ray Practices: Consequences and Control (March 1986, Mumbai).

of an aluminum filter of a millimetre thickness which will remove all the unwanted soft X-rays from the X-ray beam. This helps reduce the dose to the skin; I explained that these X-rays do not reach the film. Generally, such added filters are always in place to harden the X-ray beam. Occasionally, the radiographers remove them so that they can position the patient correctly using the light beam provided by the X-ray unit. Often, they do not know the technical needs of the filters; they do not care to put the filter back into position. My explanation led to more questions. This mini "viva voce" helped me later! Those who worked closely with Ramanna must have noticed his enviable curiosity in all science-related matters.

Dr. Ramanna delivered the keynote address and inaugurated the seminar. He urged the delegates to enforce steps to make X-ray installations safe. The media took note of his speech.

After a few days, Dr. Ramanna called a smaller group of senior scientists to discuss the policy matters associated with the enforcement of X-ray safety. Besides Prof. A. K. De, Dr. M. R. Srinivasan, Dr. P. K. Iyengar and Shri N. Srinivasan among others attended the meeting. While I spoke on the legal and practical aspects of the issue, a prominent member of the group went into "a promotion interview" mode and started to skin me alive! Dr. Ramanna, who knew the nuances of the subject, defended me. This small gesture on his part made a profound impression on me.[3]

Ramanna believed that over-regulation may be as bad as no regulation. He felt that in regulatory matters, evolution is better than revolution. He desired to strengthen AERB legally to consolidate its activities and to further its status as an effective regulatory agency.

Interestingly, in 1997, Dr. Ramanna chaired a committee – the Raja Ramanna Committee – constituted to review all aspects of the regulatory process of nuclear installations. In August 1997, the Raja Ramanna Committee submitted its recommendations. It recommended the amendment of the Atomic Energy Act 1962 to increase its effectiveness in the regulation of nuclear safety and changes in the regulatory systems so that it becomes more effective.

Ramanna's Critical Analysis of a Poignant Event

An unfortunate event reveals yet another facet of Dr. Ramanna's character. A servant murdered a scientist under gruesome circumstances.

She was my batch mate. She was a brilliant scientist, a gold medalist and a Ph.D. from a prestigious American University. She was pregnant. Since she and her husband were working, they had engaged a servant for domestic help.

Being highly educated and cultured, they freely mixed with the servant. He mistook it. On the fateful day, when her husband went to work, the servant tried to misbehave with her. She resisted valiantly. The servant stabbed her to death. Police arrested the culprit.

A day later, Dr. Ramanna addressed a condolence meeting. Even when there was inconsolable grief, Dr. Ramanna analysed the poignant event, clinically, rationally and objectively, for others' benefit.

Ramanna *vs.* Seshan on Environment: Hilarious Banter

At the start of a meeting in Old Yacht Club, Dr. Ramanna referred humorously to some "serious discussion" he had with Shri T. N. Seshan on the "deteriorating environment" in Mumbai. Shri Seshan has just then taken

---

[3] Ramanna: A Doyen among Scientists, *The Hindu*, 30 September 2004.

charge of the Ministry of Environment and Forests. They travelled together from Delhi on an early morning flight. When they landed in Mumbai, Seshan's car did not turn up at the airport in time because of some communication gap. Dr. Ramanna offered him a lift.

After coming out of the airport, a few minutes later, Ramanna noticed the row of people on either side of the road, with pitchers, water-filled cans, etc. in hand.

"Seshan, in your department, who is in charge of taking care of this deteriorating environment?" Ramanna asked. He did not reveal in what way Seshan reacted. Apparently, he was tongue-tied. Seshan was known for one-liners! (Famous comment, I eat politicians for breakfast!). The fact is that no one is bothered.

Ramanna, the Patriot

Ramanna was intensely patriotic. When he spoke in the Parliament, MPs listened to him with rapt attention. Wherever he served, he left an indelible imprint.

As a person who spurned the charm of greener pastures and responded to the call of Dr. Bhabha to come back to India, he was concerned about brain drain. He felt that the Training School churned out scientists for the future and helped greatly to stall "the emigration syndrome". Dr. Ramanna's contribution to the Training School programme is as significant as his role in placing India on the nuclear map of the world.

Dr. K. S. Parthasarathy is a former Secretary of the Atomic Energy Regulatory Board and a former Raja Ramanna Fellow, DAE. He was a trainee in the 7th Batch (1963–1964) of the Atomic Energy Establishment Trombay Training School. He secured an M.Sc. in Physics from the University of Kerala and a Ph.D. from the University of Leeds, UK. He specialised in Medical Physics and Radiation Safety including Regulatory Aspects. He was attached to the UK Medical Research Council's (MRC's) erstwhile Environmental Radiation Research Unit as a Colombo Plan Study Fellow. He was a Research Associate at the University of Virginia Medical Centre, Charlottesville, USA.

Chapter 10

Harnessing the Benefits of Nuclear Applications through the IAEA: Dr. Raja Ramanna's Roles and Expectations

N. Ramamoorthy

National Institute of Advanced Studies, Bengaluru, India

nramasta@gmail.com

Preamble

The Indian Atomic Energy Programme and the International Atomic Energy Agency (IAEA) have close association since the inception of the IAEA in 1957. The vital roles played by Dr. Homi Jehangir Bhabha, especially preceding the formation of the IAEA as well as during the early years of IAEA, are very well recognised. Dr. Raja Ramanna was one of the young colleagues involved by Dr. Bhabha since the early years with IAEA matters and events. Much later, as the Director of the Bhabha Atomic Research Centre (BARC) and subsequently as the Secretary of the Department of Atomic Energy (DAE) and Chairman of the Atomic Energy Commission of India, Dr. Ramanna had direct interactions with and contributions to the IAEA programmes, technical events and projects and IAEA Annual General Conferences (GC).

Dr. Ramanna contributed an article entitled "Positive aspects of the work of the IAEA" to the commemorative volume of 40 years of the

IAEA, called "IAEA – Personal Reflections" published in 1997. In this article, Dr. Ramanna had shared his impressions and analysis of the IAEA functions and the connection to many nations' nuclear science and technology (S&T) programmes including that of India [1]. He cited the positive progress, as well as the political impact faced, in certain areas of the IAEA functions. Furthermore, in his Autobiography *Years of Pilgrimage* published earlier in March 1991, Dr. Ramanna had commented on the IAEA functions and his interactions with the IAEA [2]. This chapter is accordingly based largely on Dr. Ramanna's own narration and supplemented by the author's experience of working at BARC/ DAE and IAEA [1–3].

IAEA and Its Origin

The Atoms for Peace address was delivered to the United Nations General Assembly (UNGA) on 8 December 1953, by US President Eisenhower. This was the culmination of numerous approaches and dialogues attempted (not always successfully though) in the face of USA–USSR policy stand on nuclear materials, technologies and associated proliferation issues. Subsequent efforts led to the UNGA approving on 4 December 1954, the creation of a new global entity for atomic energy. A series of further steps over the next year, involving select nations, led to the 1955 Conference on the Peaceful Uses of Atomic Energy (C-PUAE) held in Geneva, with Dr. Bhabha as the elected President. Dr. Ramanna was a member of the Indian delegation at C-PUAE. The 1955 Geneva Conference was the starting point for the dissemination of progress made in the development of atomic energy pursuit since much of the subject was until then shrouded in secrecy. The more advanced nations were, however, keen to know about one another's developments. Information was shared at the C-PUAE even on the projects handled during the Second World War and shrouded in secrecy until then.

The developing countries looked forward to their national development programmes receiving the benefits of using nuclear technology including the potential of nuclear electricity. A considerable amount of nuclear-related information was shared at the C-PUAE. By the end of the C-PUAE, the general feeling was that all interested countries should participate in the peaceful use of atomic energy in some organised way and not have to repeat what had been done by another country.

The sequential events of 1954–1956 formed a classic case of diplomats among the political representatives and (nuclear) scientists coming together to harness the fruits of nuclear technologies and seeking a specific entity for nuclear matters in the UN system, and at the same time by addressing the sensitivity, concerns and efforts to hold control by the key stakeholder nations. This progress eventually led to the conceptualisation and formation of the IAEA in 1957, after evolving an acceptable Statute. The Statute was the product of again very intense negotiations among political representatives, diplomats and scientists, and all this was under the scrutiny of "powerful nations" involved.

One of the drivers for the creation of the IAEA was to bring national governments together, give guidance on nuclear technology matters in different fields and collate information, be it for electricity production, or the use of isotopes in medicine and agriculture, or the basic and life sciences. Accordingly, the work of the IAEA began to get organised and gave a structure to international nuclear collaboration (cf. mostly bilateral initiatives between some countries in earlier times). India is a founder member of the IAEA and has been on the Board of Governors (BoG) of IAEA since its inception, as one of the designated members.

Scientific Advisory Committee (SAC)

The science-diplomacy synergy approach was a crucial factor in the establishment of the IAEA. This approach remained the strong foundation for its continual growth, implementing various programmes and delivering benefits to its Member States (MS), and more importantly, addressing (inevitable) political pressures due to the very nature of nuclear programme.

In this context, the IAEA's Scientific Advisory Committee (SAC) played an independent, pivotal role in programme planning and implementation in the first three decades. This was a strategic mechanism instituted in 1958 and nurtured as the means to avail inputs of leading scientists and science managers of many nations (including India) in the management of the programmes and functions of the IAEA. The first SAC created in November 1958 comprised seven esteemed members (Dr. Bhabha was one of them[1]), who were scientists of outstanding global stature. During 1983–1987, Dr. Ramanna was the Chair of the SAC including for its final

---

[1]Dr. Bhabha was on his way to Vienna to attend the 14th Meeting of the SAC at IAEA, when the untimely end to his life came due to his flight crashing or vanishing over the Alps on 24 January 1966, killing all the occupants.

(36th) meeting held in November 1987. The role of the SAC was to address the aspirations of the developing nations while enabling the pursuit of the larger goals of nuclear energy generation and ensuring keeping an eye on the long-term programme planning needs of the IAEA. Dissemination of knowledge and information sharing formed a key feature of IAEA's programme planning. It goes to the credit of science-driven approach launched through the SAC striving to counter any political impact that enabled the IAEA to not only contribute well but also gradually raise itself as the primary global nuclear forum.

It is worth recalling Dr. Ramanna's own narration on SAC: *The early years of the IAEA were guided by a Scientific Advisory Committee (SAC) which had some of the most distinguished scientists of the world and it is this guidance that made the quality of work and publications of the Agency of the highest order. Besides Bhabha, I recall the names of I. I. Rabi, John Cockcroft, W. Lewis, Bertrand Goldschmidt and others. The IAEA's own laboratory at Seibersdorf, working in collaboration with many well-known laboratories around the world, has contributed to the required data with all the necessary quality and accuracy. This would not have been possible without the help of the SAC. For reasons not clear to me the Agency decided to discontinue SAC, and I have the feeling that several countries did not like the way SAC took independent decisions which were outside the policy of the more influential countries concerned. I was, therefore, the last Chairman of SAC.*

The term of the SAC was not continued after 1987. It was only well after a decade that the then Director General (DG), Mr. M. ElBaradei, restored a similar mechanism for availing MS experts' advisory inputs, this time at IAEA's major programme level (cf. SAC's holistic coverage of all the programmes). Based on an in-depth review done in 1998 by a Senior Experts Group (SEG) established by DG, Mr. M. ElBaradei (Dr. R. Chidambaram was a Member of SEG), the Standing Advisory Group for Nuclear Applications (SAGNA) and for nuclear energy (SAGNE) was formed by the IAEA-DG. These high-level advisory committees have been functioning since 2000. The earlier established International Safety Advisory Group (INSAG) was already rendering advisory guidance to IAEA's nuclear and radiation safety programmes from 1987.

It is an interesting coincidence that in 2000, Dr. Ramanna's younger colleague Dr. V. S. Ramamurthy, as Secretary, Department of Science and Technology, India (later Director, NIAS), was chosen by Mr. ElBaradei to be the first Chairman of SAGNA. He held that position for nearly a decade from 2000, directly advising, reviewing and guiding the IAEA programme planning of nuclear science and associated applications and their interface with nuclear energy programme. The contents of IAEA's annual reports of this period contain evidence to show how much of Dr. Ramanna's expectations have been addressed by the IAEA.

IAEA Functions and Geopolitical Impact

Over the years, the atomic energy programme crossed its early phase of progress from the many national laboratories and moved on to covering large industrial-scale operations, including involving the safety of the nuclear plants and environment. The topics included addressing various techno-economic matters involving large financial investments, (national) policies and nuclear-related political issues. Naturally, there was a consequential change in the accent of the programme functions of the IAEA.

The capability to use atomic energy was considered an important part of development by many countries. The two major accidents in the field, however, impacted the interest and the IAEA activities. The first accident was in the most advanced country, USA, in March 1979 – Three Mile Island (TMI) accident; the second was in April 1986 at the Chernobyl Nuclear Power Plant (NPP) in another advanced country, the erstwhile USSR (Ukraine). The latter disaster was not only too severe a catastrophe to deal with but also caused immense psychological damage to people or societies everywhere.

Dr. Ramanna felt that these accidents were unduly used by the *press and motivated groups* to cause much harm to the pursuit of *the only source of* (clean) *power of the future*. Apart from these two accidents (ed. till then), and more importantly, the close connection between nuclear energy and its weapon potential for mass destruction *cum* military use, invariably impacted many national policies and the IAEA. Thus, Dr. Ramanna highlighted the time when the political aspect of the work of the IAEA strongly centred on the question of diversion of fuel designated for the purposes of power production to weapons development and that safeguards on fuel becoming the major topic for discussion.

The functional philosophy of the IAEA in the first phase period involved training of personnel, exchange of technological information, prototype plants, new materials of construction, etc. The use of isotopes in industry, agriculture and medicine, basic nuclear sciences and the dissemination of information through conferences and allied meetings continued to remain attractive and relevant for many Ms. Dr. Ramanna noted that it cannot be however the main programme of the IAEA for all time, although international collaboration among interested countries in these activities should be further supported. It was his expectation that at the same time, acquiring and operating nuclear power plants and associated facilities and addressing associated safety aspects would be the more important topics to be taken up and more prominently than before.

It was thus Dr. Ramanna's expectation that the second phase should include the standardisation of the design of nuclear components, along with greater involvement of the private sector utilities. He also realised that the demand for fuel and its security would always be a problem. The use of fertile material such as thorium should hence be examined. This will also help simplify the many problems concerning the present process of extraction of uranium as fissile material. This was not due to fear of any sudden shortage of uranium but because supply conditions could become dependent on nations' power politics, leading to non-availability when needed.

There was also the important question of the possible use of the available fissile material in stockpiled weapons of the nuclear weapons countries (also called P-5 countries). If it is not utilised in nuclear reactors, it will remain a political problem as well as act as a very attractive tool to terrorists. In such a situation, the concerned governments will become more involved in controlling fuel movements, questioning sovereignty and (alas!) thinking of punitive measures, even as the diversion of fuel may continue. In this context, Dr. Ramanna recalled the early years of IAEA, when SAC played an important role in the work of the IAEA and decisions were taken mainly based on scientific rationality. Later, politics and economic theories began exerting a strong influence on the IAEA functions. Dr. Ramanna lamented that it is also noted that more economists and administrators are there as the national representatives in IAEA's BoG; similarly, many diplomats have replaced the nuclear scientists as delegates at the IAEA-GC Sessions.

Before concluding this section, it is pertinent to make the following observation. Well beyond 2000, nuclear power was well on its way to expansion and/or adoption in many countries and with advocacy for IAEA support too. This pursuit had a major setback when the Fukushima Daiichi accident occurred in Japan in March 2011, leading to reversal in many nations' policies on nuclear power. It also impacted the plans of the then DG, (Late) Mr. Yukiya Amano, in his 2nd year of tenure, for supporting the adoption of nuclear power. It took well over a decade to again foster advocacy of nuclear options in the energy mix as a clean energy source. IAEA's strong positive stand, led by its current DG, Mr. Raphael Grossi, in many a global forum, on energy and sustainable development in the past few years, as well as IAEA's other on-going initiatives (e.g. Milestones Approach, Integrated Nuclear Infrastructure Review (INIR), Small Modular Reactor (SMR) support, etc.), is noteworthy. This would have made Dr. Ramanna smile in appreciation and perhaps say in his customary poetic or musical style: *better late than never*!

Leveraging IAEA to Foster Collaboration and Capacity Building in Nuclear Science

Right from the early years of IAEA functioning, assistance to training, especially for developing countries (mostly under IAEA's Technical Assistance Programme), was given importance and naturally so. An example is the conference on the utilisation of reactors for peaceful purposes held in 1962 in Bangkok. Dr. Ramanna as a participant realised the 1962 Bangkok conference to be an *eye-opener*. The participants were from different countries in the region, and yet there was a common cultural heritage to unite them for potential cooperation. This topical series has since then made numerous contributions in capacity building, human resources development (HRD), etc. in many MS.

One of the aims of the IAEA was to encourage regional collaboration in the utilisation of reactors, particularly in the SE Asian countries. The USA had set up research centres with reactors in this area, but these countries did not have trained HR for basic or applied research. To address this problem, Dr. Ramanna suggested that the countries get together to jointly make use of the facilities by pooling their manpower and other resources. Manila was chosen as a centre for this purpose and to the great enthusiasm of younger scientists of the region. Funding from the USA, Japan and Australia was offered if reactor-based research was restricted to support agriculture, health and related fields. A well-knit group from nine countries managed to publish a paper in *Physical Review* on some aspects of neutron crystallography, soon after the IAEA event on the research utilisation of reactors. The countries under the co-ordination of the IAEA had notable success, e.g. in agriculture by producing new mutants, studying materials using neutrons, use of isotopes in many ways and producing safety manuals. Yet, it was apparent to Dr. Ramanna that the funding countries cherished suspicions about the intent and capabilities of the countries coming together!

Thanks to Dr. Ramanna's support and conviction in regional cooperation under the IAEA, India (BARC) played a key role in the establishment of a new initiative in 1972, a scheme called "Regional Cooperation Agreement for Research, Development and Training Related to Nuclear Science and Technology for Asia and the Pacific (RCA)". This scheme, which evolved from the earlier success of the tri-partite agreement among India, the Philippines and the IAEA, was meant to facilitate and foster regional cooperation under the auspices of the IAEA. Dr. Ramanna assigned the charge to Dr. V. K. Iya, as the leader of the radioisotope pro-gramme of BARC, to handle the initiative to its fruition and made him the first RCA National Representative of India. RCA has been a flagship programme for over 5 decades and currently has 22 State Parties. RCA has considerable achievements to showcase, e.g. in the areas of industrial, environmental, medical and food and agricultural applications and more importantly for HRD. With IAEA support, the RCA has promoted HRD in the region. More than 170 RCA projects have enabled training for 10,000 counterparts through 650+ regional training courses. More than 4,500 experts have been engaged to provide expertise, experience and skills for the safe, effective and efficient use of nuclear technology in sup-port of sustainable development.[2] Similar cooperation schemes covering other regions of the world came up much later under the IAEA (e.g. ARCAL for countries of Latin America and AFRA for countries in Africa). The regional cooperation approach is thus yet another credit attributable to the vision and early institution of appropriate mechanisms by Dr. Ramanna and Dr. Iya.

Around 1980, India (like a few developed countries) opted to volun-tarily stop being a recipient of IAEA's Technical Assistance support (later called Technical Cooperation (TC)) so that it could be used for countries which had a greater need and participate only as a donor. This was also intended to counter the efforts of certain countries to place (undue) pre-conditions (due to the 1974 peaceful nuclear explosion (PNE) experiment

---

[2] https://www.iaea.org/about/partnerships/regional/cooperative-agreements/regional-cooperative-agreement-for-research-development-and-training-related-to-nuclear-science-and-technology-for-asia-and-the-pacific-rca, https://www.iaea.org/newscenter/news/rca-recognizes-key-achievements-with-awards-on-50th-anniversary.

done by India) for receiving TC support. Later, in 1997, in a talk, Mr. Hans Blix, DG of IAEA (1981–1997), at a Seminar in Mumbai, referred to India's voluntary action of foregoing the TC benefit and nudged to reconsider the same, by saying,[3] *"Personally I think the reasons which once led India not to be a recipient of the (TC) programme have lost their relevance. Just as India contributes to the programme, I think India could benefit from it".* Yet, India, as a matter of its principled policy, has been participating only as a donor in IAEA's TC programme.

IAEA provided an apt forum for many cooperative pursuits including addressing the interests of advanced countries. One such case was the interest of Norway, with Dr. Ramanna serving as the Chairman of an IAEA Committee to supervise the programme on developing a small heavy water-based research reactor (called NORA project, steered by Norway). This project showed how the IAEA forum fostered bringing together global talents and leadership to effectively address relevant topics of contemporary interest to one or more MS, including advanced countries in the nuclear field.

Similarly, much later, a project on new reactor technologies for power production and associated fuel cycle aspects called the International Project on Innovative Nuclear Reactors and Fuel Cycles (INPRO)[4] was anchored by the IAEA to the benefit of many countries interested in nuclear power. India has been an active member of INPRO since its inception in 2000. Dr. Ramanna's advocacy of the need for focus on thorium utilisation has been the key aspect of India's contributions to INPRO.

IAEA, with its worldwide reputation, continually adjusted with the times by tailoring its programmes on what is most useful or relevant to its MS. After decades of existence, certain problems of earlier years have been naturally (re)solved or, as it often happens in science, have become obsolete. On the other hand, addressing more pressing aspects was given priority to make IAEA deliverables more relevant to contemporary needs. Other strategies were also adopted to enhance the deliverables, for example, coordinated actions with WHO for health care applications, with

---

[3] https://www.iaea.org/newscenter/statements/overview-international-scenario-nuclear-energy-and-its-role-sustaining-development.

[4] https://www.iaea.org/services/key-programmes/international-project-on-innovative-nuclear-reactors-and-fuel-cycles-inpro.

NEA-OECD for nuclear energy and fuel cycle topics, as well as with other relevant professional entities.

Member or Leader of Indian Delegation to IAEA-GC

Dr. Ramanna attended the IAEA Annual GC as a member of the Indian Delegation for several years, starting from the tenure of Dr. Vikram Sarabhai as Secretary, DAE and Chairman, AEC and then when holding the post of Director-BARC (1972–1978 and 1981–1983). Further, during 1983–1986, as Secretary, DAE and Chairman, AEC, Dr. Ramanna was the leader of the Indian Delegation to the IAEA-GC. Thus, he directly witnessed the heat of the global reactions to India's PNE of May 1974 included in certain key national statements to the IAEA-GC! It would have been even more striking for him, as these key national teams knew his main role and leadership to PNE!

It is pertinent to refer to Dr. Ramanna's observations on PNE, IAEA, in this context. Under the IAEA, seminars on the peaceful uses of nuclear explosions (PNEs) had been organised in the past. Before 1974, there was much publicity given to the many possibilities of PNE's potential for certain explosive operations, which normally would require considerable time, or maybe even impractical for one or more reasons. Advanced countries had done many demonstrations to show such possibilities. To witness such a demonstration, a group of scientists from different countries was even taken to a PNE site in the USA. That experiment was known as "Project Rulison". Alas, after the Indian PNE test in May 1974, the above views got reversed on such potential benefits of PNEs and they were branded as *useless and far too expensive*. At the IAEA seminar on the subject, the latter view was projected, mostly by economists and diplomats (cf. scientists, who informally shared different ideas).

Much later, the global reaction and response to the April 1986 Chernobyl accident were witnessed by Dr. Ramanna at the 1986 GC. Some of the GC resolutions that year naturally called for concerted actions to support strengthening the safety of nuclear power plants, addressing trans-boundary impact and emergency preparedness, etc.

Furthermore, by sheer coincidence, it was India's turn in 1986 to be the Head of the IAEA-GC! Dr. Ramanna, as the leader of the Indian

Delegation, got the privilege of serving as the elected President of the IAEA-GC in September 1986 held at the imperial Hofburg Palace in Vienna. The GC of 1986 turned out to be the last GC held at that prestigious royal venue in Vienna.

Dr. Ramanna's observations on the approach and systems for safeguards of nuclear material indicated how much discriminatory tactics and positions were taken by advanced countries and exerted pressure on developing countries planning to embark on or expand their nuclear programme. He greatly regretted the political impact, including on the IAEA and on its GC Sessions. His own narration runs along the following lines. In the next phase of the international scenario, debates on safeguarding nuclear materials dominated. It was evident that the advanced countries were looking upon the issue as another opportunity to strengthen their stranglehold over the other (underdeveloped) countries. Developing nations desired that inspection should apply to all nations using atomic energy. The advanced countries ignored the suggestion and would only have their own inspection systems! Eventually, several countries had to buckle under pressure.

Musical Benefits of Visits to Vienna

Dr. Ramanna visited Vienna (IAEA) frequently since 1960, particularly for the SAC meetings and the IAEA-GC. Dr. Ramanna was a passionate lover of Western classical music and a gifted pianist. His well-acclaimed interest and knowledge in Western music received an additional boost during his Vienna visits since Vienna has been the centre of Western classical music for more than two centuries with an impressive musical history. He has cited it as a great bonus from his visits. He mentioned as an example the renowned opera in Vienna set in the past splendour which he could visit to his heart's delight (despite *exorbitant rates charged for the Vienna Opera show – perhaps to extort money from tourists*, he said!).

Dr. Ramanna was an extremely ardent admirer and devotee of the great music composer Mr. Franz Liszt of Hungary (1811–1886). In 1986, on the death centenary of Mr. Liszt, Dr. Ramanna gave a piano concert at Kamani Theatre, Delhi. This led to the Embassy of Hungary, and GDR (East Germany), approaching him with much appreciation. In turn, Dr. Ramanna made use of the above embassies' support to facilitate his visit to Budapest and Weimar (during his visit to the IAEA for the SAC meeting). He visited Liszt Museum in Budapest accompanied by the Indian Ambassador in Vienna. His subsequent visit to Weimar (via East Berlin) took him to Liszt Academy and meet the music professors there (*"some of the best musicians of Europe"*, said Dr. Ramanna; he was asked to play for them, but he said that he could not perform anywhere near to his best!) and then go to Liszt's house, where he could see, and even touch with reverence, the great composer's piano standing in its original location! Furthermore, an extremely able pianist in the group played select works of Mr. Liszt requested by Dr. Ramanna! No wonder Dr. Ramanna called it his *Liszt Pilgrimage*!

Concluding Remarks

India has continued to remain actively engaged in all the key programmes of the IAEA leading to enhancing its global contributions. Programme interests and synergies have been well harnessed to mutual benefits. Some of the apprehensions cited in the 1997 article of Dr. Ramanna can be recalled, whenever any apparent conflict is seen between the views of certain countries *vis-à-vis* the needs or interests of several developing countries. Yet, Dr. Ramanna acknowledged, to the credit of the IAEA, that *"of all the UN organizations, the IAEA has been the most effective in the implementation of its programmes"*.

Dr. Ramanna was not alive to see the IAEA and its DG Mr. Mohamed ElBaradei being awarded the Nobel Peace Prize in 2005. He would have been surely delighted and recalled his own roles for the IAEA over four decades starting from its inception in 1957.

Furthermore, he was not around to see the progress in India achieving civil nuclear cooperation agreements with select nations which also involved a well-negotiated specific safeguard agreement for India with the IAEA accomplished in 2008. Despite the political stand and pressures from certain countries, India and the IAEA entering into such an agreement on safeguards would have been unimaginable, including perhaps Dr. Ramanna, given his past experience and observations on the political impact on the IAEA by certain countries.

It is a matter of sheer coincidence that Dr. Ramanna breathed his last in a Mumbai hospital in the same week when the 2004 IAEA-GC was underway in Austria Centre, Vienna.

Further Reading

D. Fischer, *History of the IAEA: The First Forty Years: A Fortieth Anniversary Publication*, IAEA/STI/PUB/1032. IAEA, Vienna, 1997.

IAEA – Personal Reflections, A Fortieth Anniversary Publication, IAEA/STI/PUB/1033. IAEA, Vienna, 1997.

www.iaea.org, www.dae.gov.in, www.barc.gov.in.

R. Ramanna, *The Structure of Music in Raga and Western Systems*. Bharatiya Vidya Bhavan, Bombay, 1993.

References

1. R. Ramanna, Positive aspects of the work of the International Atomic Energy Agency, in *IAEA – Personal Reflections, A Fortieth Anniversary Publication*, IAEA/STI/PUB/1033, 1997, pp. 287–295.
2. R. Ramanna, *Years of Pilgrimage – An Autobiography*. Viking, India, 1991.
3. N. Ramamoorthy, *et al.*, Harnessing Science – Diplomacy Synergy (SDS): Key Feature of the IAEA Roles and Success since Inception (INIS-XA-24M3947). International Atomic Energy Agency (IAEA), November 2024. https://inis.iaea.org/search/55101626. INIS, 2024. IAEA, Vienna (in Press).

Dr. N. Ramamoorthy has been an Hon. Adjunct Professor at NIAS, Bangalore, since 2016 and is handling simultaneously advisory or consultancy tasks for AERB, Mumbai, and IAEA, Vienna. Earlier, he held senior managerial positions at BARC, BRIT and DAE and also served as Director of a Division at IAEA for 7.5 years. He has over 45 years of professional and managerial experience in the fields of production and utilisation of radio-isotopes and radiopharmaceuticals, radiation technology applications and associated regulatory matters of safety and security. He is the co-author of the book *Fundamentals of Radiochemistry* of IANCAS, Mumbai, co-editor of the book, Ionising Radiation and Mankind, of Cambridge Scholars Publishing, UK, and editor of two compilations, *Advanced Radiation Technology*, of World Nuclear University, London (2019), and *Knowledge Management and HRD Applied to Radiation Technologies*, of Rosatom Technical Academy, Moscow and IAEA, Vienna (2021).

Chapter 11

Dr. Raja Ramanna: My Short Associations – Lasting Impressions

Subramania Jayaraman

Retired Clinical Radiotherapy Physicist

Arizona, USA

jayaraman_phoenix@yahoo.com

First, I would like to mention that I have not closely worked with Dr. Raja Ramanna to the extent to say that we knew each other. The narratives that follow are merely some incidences where we came in touch with each other.

I joined Atomic Energy Establishment (now the Bhabha Atomic Research Centre (BARC)) Training School in August 1963 as a 19-year-old. On the day of the inauguration, Dr. Ramanna gave a pep talk to all the trainees. He mentioned Dr. Homi Bhabha's belief in finding scientific talents and building research labs around them. During the year, he visited and addressed us a few more times. He asked us to strive for high scores even if 50% and 60% were indicated as two levels of passing grades. The higher the rank, he said, the better the chance to get into a superior group, such as Nuclear Physics, Theoretical Physics and Astro Physics. Otherwise, one was likely to get assigned to a group (of lesser glamour) such as Health Physics. He seemed to sound a bit harsh then, yet he was only motivating us to do better and not to coast to the end and take the study easy.

The training ended in July 1964. My friend – nay, my guide in those young years of mine – V. S. Ramamurthy (VSR), got the top rank. He preferred to work with Dr. Ramanna and opted for Fission Physics. When I was called to tell my preference, even before I said anything, Dr. Ramanna, who was conducting the assignments, said, "You do not need to make a choice and then they also need to want you or accept you. You are someone in demand already somewhere. You will do well there". I was assigned to work in Radiation Applications and Related Safety. I decided I would accept the assignment with grace and just do my best there. My division had the name Directorate of Radiation Protection (DRP). My friend VSR suggested that, within DRP, I select to work in the medical use of radiation. That was the good beginning of my long career up to age 67.

I had completed three and a half years of service at DRP. Then, at the age of 23, I was called in by my Division Head, Mr. P. N. Krishnamoorthy. He asked me to fill out a form for a job application at the International Atomic Energy Agency (IAEA), Vienna, Austria, which I did. He guided me as I filled in the form. I was accepted for the job in Vienna and went to serve there for two years. There, I helped continue a dosimetry monitoring service to radiation therapy centres around the world. The dosemeters that went out from Vienna were irradiated and returned with details of their dose derivation. The various ratios and factors used for the dose estimate were examined and doses recalculated at IAEA.

The errors in individual factors can be small – at 1–3% levels. Yet, they give errors of 5–10% in combination, which can affect the clinical outcome. The recalculations were reported back to the individual clinics. They were invited to participate again in the next cycle. That dosimetry service from IAEA has continued with possible improvements to this day.

In Vienna, I learned to drive and owned a used Volkswagen car in good condition. I could drive comfortably around. Once, Dr. Ramanna came as a part of a high-level delegation from India. I introduced myself to him and expressed an interest in meeting him in his hotel room, to which he agreed. When I met him, I asked him whether I could drive him around to show the beautiful architecture of buildings and gardens on the Inner Ring Road or visit some department stores. He wanted to shop for a new briefcase of an appropriate size to meet his needs. We visited the three major departmental stores. He selected a briefcase of his liking after verifying that it was of the right dimensions to accommodate things he would carry. We used items from the stores, placed them inside and checked. We went to return the items to their respective places within that store. Then, while passing the kitchen department, I asked him, "How about buying something for your wife's kitchen?" He, very slowly, responded, "Ok, Krishnan". In the kitchen section, based on my experience with others, I recommended some small tools such as potato peeler and vegetable shredder. He selected some of them. Time and again, he addressed me as Krishnan. I corrected him just once. His trying to address me by name meant that he was feeling some closeness and affection towards me as a person. Years later, I came to know about his deep interest in Mukunda Mala of Kulasekara Azhwar on Lord Krishna. Except for those shopping moments, he was taken care of by the local Indian embassy throughout his stay.

I finished my two-year work in Vienna and returned to BARC on 1 January 1970. Some seniors with whom I worked at IAEA, Vienna, were continuing there. Some had gone back to their respective countries as I had done too. In the year 1973, I received an invitation from the IAEA Section where I had served to attend a three-day meeting in Vienna as a consultant to formulate the contents of a newly planned IAEA publication. I submitted the letter to the then Division Head, Dr. K. G. Vohra. He, in turn, submitted it to the Trombay Scientific Council (TSC) Meeting and came with the approval for me to go to Vienna. I attended the meeting and returned. I was called upon to make a presentation to the TSC about my meeting. I said in my statement, "The essence of the meeting, as I understood, was that the use of radium sources in medical clinics must end for radiation safety reasons. Radium is to be replaced with reactor produced substitutes like Cs-137, Co-60, Ir-192 etc. We are already doing that here using Co-60 and Ir-192. However, for gynecological applications Cs-137 has become the preferred option more than Co-60. We should also produce Cs-137 sources from fission product extracts". Dr. Ramanna was presiding. He smiled and looked like he was relishing what I had just said. But he stopped me and remarked, "Oh, are you not getting carried away? We will listen to all of that later! O.K.".

I think it was around 1973–1976. Anushakthi Nagar, where apartment buildings were coming up for allotment to staff, was rapidly growing. There a lot of buildings were being completed, and flats were being allotted to staff adopting a policy that combined service seniority with monthly income, for eligibility. The trainees from the 7th batch felt that there was some anomaly in that. The buildings were constructed from low categories first to higher categories later. A group of us who joined as Trainees

from the 7th batch were always eligible for yet-to-be-built higher categories. As we moved up the income/career ladder, we were migrating to
even higher categories and newer waiting lists. Thus, it looked as if
we would forever be on the waiting list for the buildings yet to be built.
I thought it would be appropriate to draw this fact to the attention of the
authorities. I took it upon myself to speak about this to Dr. Ramanna.
After all, I felt he had been our "caretaker" from our Training School
days. I asked for an appointment to see him on behalf of the 7th batch
trainees. I was called to meet him. As I entered his office, he offered
me a seat. I promised him that I would be brief. He did not seem to
remember me from my previous occasions with him. I explained the
situation to him. He let me talk with no interruption from him. I could
completely state the case. He responded by directing me to speak to the
individual members of the Trombay Council (TC) and explain the same
to them. I was astonished by Dr. Ramanna's capacity to give such complete listening to a distant subordinate like me. The TC consisted of
Directors of Groups of Scientific Divisions and the Administrative
Controller. I was accompanied by two others from the 7th batch. I remember one of them was Dr. K. S. Parthasarathy, who later became Deputy
Director of Atomic Energy Regulatory Board (ARRB). The other was
Dr. T. K. Balasubramanian, later Head of Spectroscopy Division. The
three of us met each member of TC and presented the problem. Fairly,
our situation got resolved by a special allotment.

It was late 1975. I received a letter from the IAEA group inviting me
to serve them for a brief period of four months without giving me any
details of the work. The letter came, addressed to me, at my official
address. I presented the letter to my division head Dr. K. G. Vohra. He
presented the letter, in turn for approval, at the next TSC meeting. He
returned from the meeting and told me that the approval was not given.
He said TSC held the view that any such letter should not have come
directly to the person in question.

It should have come through the higher channels! I felt that something
was not right there, as I did not solicit the invitation. I sought an appointment to speak to Dr. Ramanna and meet him. I told him that my case was
discussed at the last TSC meeting, and Dr. Vohra told me that the fact that
I received that invitation directly as posted to me should not have happened. I explained to Dr. Ramanna that any letter posted to my address got
delivered by the postman to me. After I knew the contents of it, I gave it
to my Division Head, who presented it at TSC. I was the one subjected
to the rules and regulations of my institution and not the staff in Vienna.

But it seemed that there had been a presumption that I might have sought the invitation. I wanted to clarify that it was not so. He allowed me to complete what I needed to communicate in a few sentences. He heard me out fully without interrupting or cross-examining. I found him to be so very graceful that such a top-senior person would give me a full audience. After this, about a six-week period went by. I had totally forgotten about this. One day Dr. Vohra called me and said, "Your case has been approved". I could not right away connect to what he was referring to. The same item had been rediscussed at TSC, and this time found approval. Following that, I went to Vienna for about five months. During this trip, I was called to evaluate the effectiveness of the program which I had helped set up in the initial stages. Many clinics that had poor results improved their calculations and had shown better results in subsequent trials. This study resulted in a scientific publication. Such quality control is now used widely around the world with appropriate local modifications.

I completed my Ph.D., in 1979, under the guidance of Dr. R. Chidambaram. I owe my confidence and growth to the trust and expectations of him and others such as Shri P. N. Krishnamoorthy, Dr. K. G. Vohra, Shri S. Somasundaram and many others in DRP and BARC. In fact, I can fill a long list of names.

I desired to widen my background by working in clinical settings treating cancer patients. Towards that end, I left DRP, BARC, in 1980 to join the University of Virginia Medical Center, Charlottesville, VA, USA. I did well in my career as a Clinical Radiotherapy Physicist till my retirement at age 67 in 2011. I could even publish a textbook *Clinical Radiotherapy Physics*, based on my own clinical experiences. It found its place in many libraries around the world but is crying out for an updated edition by competent people. In main, I would like to state that the foundation for my doing well came from BARC.

The occasions I had interacted with Dr. Ramanna had been few, short and far between. Yet, when I look back, they have been memorable and fruitful for me. He was trustful and expecting, inspiring and influential, demanding and encouraging, endearing and considerate, listening and quite humble. I thank the editors of this volume for giving me this opportunity to express my appreciation for Dr. Ramanna.

Dr. Subramania Jayaraman was a Trainee of the 7th batch (1963–1964) of the (BARC) Training School. He worked at the Division of Radiation Protection, from 1964 to 1980 and left to work in the USA in March 1980. He had a long career in Radiotherapy Physics and worked at the University of Virginia Medical Center, Charlottesville, Rush Presbyterian Medical Center, Rush University, Chicago, Albert Einstein Medical Center, Philadelphia, Temple University, Philadelphia, Valley Radiation Oncology, Phoenix, and Ironwood Cancer Research Center, Chandler, finally retiring in 2011.

He is a recipient of the Distinguished Medical Physics Award, Indo-American Society of Medical Physics (2006), and Prof. Lawrence H. Lanzl Lecture Award, Midwest Chapter of American Association of Physicists in Medicine (1999), and has co-authored the book *Clinical Radiotherapy Physics*, with Prof. Lawrence H. Lanzl, which was published by Springer Verlag, Heidelberg, 2004 (second edition). Dr. Jayaraman has a strong interest in classical Carnatic Music sustaining him in retirement. Many kruthis sung by him are published on YouTube channels.

Chapter 12

Dr. Raja Ramanna: A Pioneer in Indian Nuclear Sciences and His Impact on Nuclear Medicine

M. G. R. Rajan

Radiation Medicine Centre, Bhabha Atomic Research Centre, Mumbai

Board of Radiation & Isotope Technology, Mumbai

mgr.rajan@gmail.com

The "Buddha Smiling" Peaceful Nuclear Experiment of 18 May 1974, was an epoch-making event for India. The names of Dr. H. N. Sethna and Dr. Raja Ramanna were in the news. It was a hot topic of discussion among my classmates in my first-year M.Sc, Chemistry course, at Central College, Bangalore. Some months later, while we were in our second-year M.Sc, Dr. Raja Ramanna visited our college and gave a talk on nuclear energy and how it has societal applications in health, industry and agriculture. There was deep admiration for this eminent scientist (with no airs). These societal benefits of nuclear energy surprised all of us. He also told us of the career opportunities at Bhabha Atomic Research Centre (BARC) and some of us planned for it.

The next time I heard Dr. Raja Ramanna was towards the end of my 21st Batch BARC Training School, in the Training School Hostel (TSH) Multipurpose Hall, where he spoke on his investigations on nuclear fission and stochastic theory of the fission process and nuclear structure. He also touched upon the various R&D activities carried out at BARC. Within a very short period, he left for Delhi where he was assigned responsibilities in the Defence Ministry and Defence Research Development Organization (DRDO).

Dr. Raja Ramanna's commitment to using nuclear energy for societal benefits, including healthcare, was apparent when he delivered the Keynote Address at the 11th Annual Conference of the Society of Nuclear Medicine, India (Delhi, November 1979). He started off by talking about consonance and dissonance in music – the former is pleasant to the ears, while the latter can be jarring that even a few dissonant notes can ruin a recital. He then drew this analogy to human health and said how it is important to identify and rectify any dissonance in human physiology before it advances to a disease. Although nuclear medicine in India was in its infancy, he expected that nuclear medicine techniques and imaging would be able to detect this health dissonance much before other modalities because nuclear medicine imaging was a truly physiological one using very specific radioactive tracers in the form of radiopharmaceuticals. Since the indigenous availability of radioisotopes/radiopharmaceuticals was still limited to the metros and state capitals, he went on to mention the limitations of CIRUS in producing the required radioisotopes and in required quantities to meet India's demands due to the neutron flux limitations and the number of ports available for irradiating targets with neutrons. He said that this problem would be over soon, as the R5, a 100-MW (thermal) research reactor, of a wholly indigenous design, was under construction and it was expected to become operational soon. It would be used for a variety of applications, including testing of equipment and materials, large-scale production of isotopes and function as a National Facility for Neutron Beam Research. It could meet most of the radioisotope requirements of nuclear medicine, agriculture and industrial applications.

Since he was then associated with the DRDO, he suggested a more active collaboration between the Institute for Nuclear Medicine and Allied Sciences (INMAS) and Radiation Medicine Centre (RMC), BARC. INMAS was established in 1964, a year after the RMC, a division of BARC, started at the Tata Memorial Hospital Building in 1963 by none other than Dr. Homi Jehangir Bhabha himself. He invited Dr. K. N. Jeejeebhoy from the UK to head RMC.

Dr. Raja Ramanna returned to BARC as Director in 1981 and hastened the completion of the R5 research reactor, which was renamed Dhruva in October 1984. Dhruva attained criticality in August 1985 and in a few years produced I-131, Mo-99, Sm-153, P-32, etc., augmenting whatever was available from CIRUS, thus ensuring that the nuclear medicine practitioners could get the required radioisotopes. The increased availability of radioisotopes enhanced the scope of nuclear medicine, from what was largely a diagnostic to including radio-nuclide therapy in a big way. The number of nuclear medicine centres in the country increased several folds.

Dr. Ramanna superannuated in 1987 as Chairman of AEC, but his legacy to nuclear medicine lives on.

The Dhruva reactor is the workhorse for radionuclide production. The weekly availability of Mo-99 (parent isotope for Tc-99m), I-131 (for diagnosis and therapy of thyrotoxicosis and thyroid cancer) and Sm-153 for bone-pain palliation have encouraged many hospitals, both in the public and private sectors, to set up nuclear medicine departments. The Radiopharmaceutical Division (RPhD), BARC, and Board of Radiation and Isotope Technology (BRIT) play an important role in facilitating this (Fig. 1).

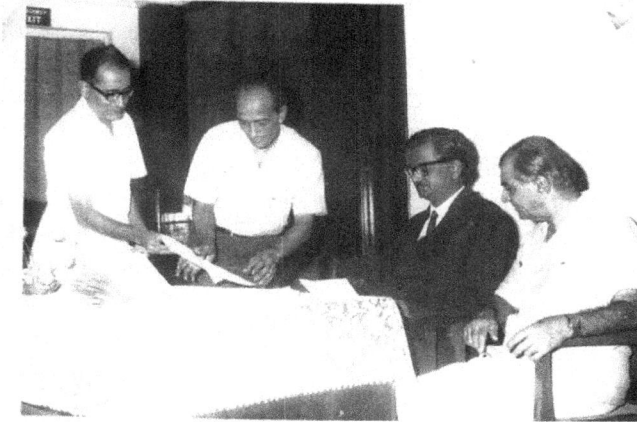

Fig. 1. (L-R) Dr. R. D. Ganatra, Head Radiation Medicine Centre, Dr. Sundaram, Dr. Raja Ramanna and Dr. Gopal Ayengar at Certificate Distribution to Students on Completion of their Short-Term Nuclear Medicine Course (1967).

The availability, since 1973, of trained nuclear medicine physicians from the Diploma in Radiation Medicine (DRM) course and nuclear medicine technologists from the Diploma in Medical Radio Isotope

Techniques (DMRIT) course has also helped the above. Dr. Ramanna provided the required institutional support to get the approvals from the Bombay University and the Medical Council of India for conducting these one-year post-graduate courses. RMC has been conducting MD (Nuclear Medicine) courses since 2915.

Lu-177 gained importance in the last two decades as a radioisotope for diagnosis and therapy of neuroendocrine tumours, prostate cancers, bone pain palliation, etc. This is achieved by coupling the isotope to different molecules. The major contribution to this is from the RPhD. Required amounts of it are produced in the Dhruva reactor and supplied from BRIT to customers as Lu-177-DOTATATE and Lu-177-PSMA for theranostics (a two-pronged approach to diagnosing and treating cancers using radiotracers) of neuroendocrine tumours and prostate tumours, respectively. The Lu-177 produced is from Lu-176 (n, γ) Lu-177, but the carrier present does not reduce its therapeutic ability. The use of Lu-177 in nuclear medicine is almost on par with I-131.

The first medical cyclotron in the country, installed in the basement of RMC and commissioned in 2002, in the TMH-Annexe building, had its beginnings in 1970 when funds for the 12-storey building were sanctioned. It was designed to provide larger premises for RMC and TMH. At that time, medical cyclotrons were in their infancy, but with the support of Dr. Ramanna, the building design, which included a large room in the basement with a 1-metre-thick hematite concrete roof, was approved and constructed. At that time, it was envisaged to get a small cyclotron to produce only fluorine-18. This provision for a vault in the basement helped in getting the necessary funds, in the IX-Plan (1997–2002), since it wasn't a green-field project. The required safety approvals to modify the room to fit a 16.5 MeV cyclotron were obtained. The cyclotron has liquid and gas targets to produce F-18, C-11, N-13 and O-15 for positron emission tomography (PET-imaging).

Since 2002, the RMC-Medical Cyclotron Facility (RMC-MCF) has been producing [F-18] fluoro-deoxyglucose [F-18] FDG and several other [F-18]-radiopharmaceuticals for in-house use as well as for supplying to other nuclear medicine centres in Mumbai. The successful and economically viable operations of the RMC-MCF encouraged many hospitals in the public sector to install medical cyclotrons. There are about 40 cyclotrons catering to nearly 500 nuclear medicine centres in the country. Incidentally, as of date, RMC-MCF is one of the two cyclotrons in the government sector that is still providing service to society.

Apart from F-18, the most used PET-radioisotope, the metallic Gallium-68, another short-lived PET-radioisotope has been found to be a very useful PET-diagnostic agent when complexed with a variety of molecules. Examples are [Ga-68]-DOTATATE and [Ga-68]-PSMA, which in combination with [Lu-177]-DOTATATE and [Lu-177]-PSMA form excellent theranostics pairs. Ga-68's parent Germanium-68 ($T_{1/2}$-270d) can be produced in high-energy cyclotrons and loaded into generator systems such as Mo-99→Tc-99m generator systems. Ge-68→Ga-68 generators are very expensive. With the installation of the 30 MeV IBA cyclotron at the Chakberia campus of Variable Energy Cyclotron Centre, Kolkata, the Ge-68 will be available indigenously. Apart from Ge-68, the 30 MeV cyclotron can also produce a wide range of radioisotopes for medical and research purposes, including 18F, 64Cu, 89Zr, 123I, 111In, 201Tl and 211At (an α-emitter).

DAE has gone a step further in approving the procurement of an IBA 70 – 250 MeV proton beam therapy machine at the Advanced Centre for Treatment Research and Education in Cancer (ACTREC), Kharghar, Navi Mumbai. This is operational from 2022. Proton therapy uses high-energy proton beams to treat cancer. The beams are more precise than X-ray beams, which allows for more energy to be used to attack cancer cells while minimising damage to healthy tissue. The more energy is from Bragg's peak that is characteristic of charged particles decelerating in matter and the energy is deposited in a very short path, minimising collateral tissue damage. The location of the Bragg peak can be controlled by adjusting the beam energy.

As a policy (formulated by Dr. Ramanna and maintained by his successors), RMC has been providing low-cost nuclear medicine diagnosis and therapy so the poor patients referred to RMC can avail high-quality treatment. The charges are a small fraction of what is charged by private hospitals. Hence, a very large number of patients come to RMC for diagnosis/treatment. This often leads to a long waiting list, particularly, when radioisotope supply is limited due to reactor shutdown/maintenance and BRIT imports some Mo-99 and I-131 to meet part of the demands. After CIRUS was shut down, the only radioisotope source was Dhruva, and the patients' waiting list for I-131 large dose therapy for thyroid cancer increased. It went up to six months in October 2013. There was a query from the Prime Minister's Office (PMO) regarding this and about action to be taken. With support from the then Director, BARC, and the BARC Safety Council, a campaign was conducted to treat the wait-listed patients. Reactor Division (Dhruva) and RPhD supported RMC and supplied the required I-131, and in the period from January to September, over 900 patients were treated and the waiting reduced to one week. The I-131 effluents from the radionuclide therapy wards into public sewerage were diluted and controlled to be within permissible levels. Since 40% of the patients coming to RMC for radionuclide therapy are from West Bengal, Orissa and the north-east region, a proposal was made to Director of BARC to start an RMC in Kolkata. This was very favourably received and funds sanctioned. Land was provided at the VECC campus at Rajarhat by the then Director of VECC. The Radiation Medicine Research Centre (RMRC) was built and dedicated to the nation on 21 November 2022. It has twice the number of beds of RMC, Parel, and is functional with positron emission tomography-computed tomography (PET-CT), Single Photon Emission-Computed Tomography (SPECT-CT) and SPECT gamma cameras.

In September 2024, RMC completed 61 years as the premier nuclear medicine department in the country rendering state-of-the-art nuclear medicine services at a very low cost. RMC is also a Teaching and R&D centre. Lest we forget this Institute, the RMC continues to fulfil the legacy of its founder, Dr. Homi Jehangir Bhabha, who started and seniors such as Dr. Raja Ramanna who have, directly or indirectly, nourished, supported and continue to support it.

Dr. M. G. R. Rajan is from the 21st Batch of the (BARC) Training School and joined the Radiation Medicine Centre (RMC) in 1978 and superannuated in June 2016 as an Outstanding Scientist & Head, RMC, BARC. He also held the adjunct post of Deputy Chief Executive of the Board of Radiation and Isotope Technology (BRIT). He was Professor at the Homi Bhabha National Institute in Life Sciences and Chemical Sciences and has guided students in their Ph.D. He specialised in the production of PET and SPECT radioisotopes using Medical Cyclotrons and converting them to GMP-approved Radiopharmaceuticals. He is also an expert on Radioimmunoassays and Related Methods and Biostatistics.

Dr. Rajan chaired the Atomic Energy Regulatory Board (AERB) constituted committee to Prepare a Safety Guide for Medical Cyclotron which was published by AERB in 2016. He was a member of the Expert Committee on Radiopharmaceuticals constituted by the Indian Pharmacopoeia Commission for preparing Monographs on Radiopharmaceuticals for the Indian Pharmacopoeia. Post superannuation, he was awarded the DAE-Raja Ramanna Fellowship (2016–2019), and he worked on the design and construction of the Radiation Medicine Research Centre (RMRC) at Rajarhat Campus of Variable Energy Cyclotron Centre, Kolkata, which was dedicated to the nation in November 2022. He was invited to provide input to DAE to submit a proposal to NITI Aayog for setting up centres of excellence for cancer hospitals and nuclear medicine centres.

Chapter 13

Raja Ramanna: Some Reminiscences

K. Kasturirangan

National Institute of Advanced Studies, Bengaluru

Indian Space Research Organisation, Bengaluru

krangank@gmail.com

I met Dr. Raja Ramanna for the first time in the most unexpected fashion. I was working at the Indian Space Research Organisation Satellite Centre (ISAC) [now U. R. Rao Satellite Centre (URSC)] and looking after the activities of the Technical Physics group. Due to the interest of Prof. U. R. Rao, the then Director of ISAC, and myself in the field of Astronomy, we were excited at the possibility of investigating the cosmic gamma-ray bursts, which were analogous to the atmospheric nuclear explosions by the researchers at the Los Alamos National Laboratory at USA. These bursts at that time were very mysterious and had very sharp rise and decay lasting for few seconds. The scientists at that time expected these bursts to come from far away galaxies, and because of the intensity of these radiations, it was conjectured that they could be the result of major explosions, sometimes outlasting temporarily the total radiation of a galaxy. Prof. Satish Dhawan, the then Chairman of Indian Space Research Organisation (ISRO), suggested, that to develop a detection system to be flown in space it would be a good strategy to discuss this with Atomic Energy Commission (AEC), who are also interested in looking at the detection of gamma rays from the atmospheric nuclear explosions. I was excited not only because of the prospect of working in this area but also at the possibility of meeting Dr. Raja Ramanna, the then Chairman of AEC and a leading scientist in Nuclear Physics. I undertook a visit to Bombay to meet Raja Ramanna based on the introduction from Prof. Dhawan. I met him at his office at the Old Yacht Club, together with Dr. P. K. Iyengar, the then Director of Bhabha Atomic Research Centre (BARC). I was quite nervous to meet such well-known luminaries at the level of AEC and BARC and that too when I was just one level above a student at that time. Raja Ramanna's office was facing the Arabian Sea on one side, and it was a very beautiful sight. This was the beginning of a fabulous association which I never expected to happen, but it did take place.

I explained to Raja Ramanna and P. K. Iyengar about our interest in detecting the cosmic gamma-ray bursts which could be from extra-terrestrial sources with huge power behind them. Raja Ramanna mentioned about the Los Alamos nuclear explosions in the atmosphere. He and Iyengar further explained about how one could build detectors with sodium iodide or caesium iodide crystals for gamma-ray burst detection. The configuration for such detection would involve suitable anti-coincidence shields to cut off the local background radiations.

Raja Ramanna expressed intense interest in cooperating with us (ISRO) in developing such a system, which could finally be used in space-craft. Raja Ramanna was quite happy with the type of discussions we had and asked me when I was heading back; I told him since there are no flights in the evening, I would be travelling the next day. He asked me to join him for a lecture on Plasma Physics, which I agreed to attend. The lecture was organised in a huge hall and Raja Ramanna who was the Atomic Energy Commission Chairman made me sit right next to him with P. K. Iyengar on the other side. All were wondering who this young chap sitting next to Ramanna was!! He then invited me to join him for lunch where many important people of AEC and BARC were present. Raja Ramanna showed such a nice attitude of familiarity towards me, who was just a youngster then, which you often don't come by in people of his cadre. This was more of an experience for me to know about Raja Ramanna as a person and the way he treated and interacted with people than learning about cosmic ray explosions. I think this first interaction with Raja Ramanna and P. K. Iyengar left an indelible impression on me regarding the kind of people one should work with to get the maximum benefit because they are top class in their understanding and are willing to share any information with you. This is one of the best experiences I had thanks to Profs. Satish Dhawan and U. R. Rao. It is appropriate to recall at this juncture that all these preparatory work with AEC and in particular with Raja Ramanna resulted in the realisation of a space mission designated as Stretched Rohini Satellite Series (SROSS) satellite, with an instrument developed by ISAC for gamma-ray burst detection. This instrument detected several gamma-ray bursts and, in a sense, became a vital component of a global patrol for detection and recording of the same.

Raja Ramanna's field was Nuclear Physics and Nuclear Energy, whereas mine was Space and areas such as Astronomy, so, there was not much of a synergy between the two, but Physics was the common binding force so a lot of things could be interpreted with Physics and that's the best thing that could happen. As I continued to study and work at ISAC, I had the advantage of knowing in the process that Raja Ramanna was working on nuclear physics, nuclear explosions, nuclear power reactor design and development and the use of nuclear programmes for a variety of areas for societal benefit whether it is related to the use of atomic energy for power, for health, etc. There was a whole host of programmes initiated by Homi Jehangir Bhabha which was continued by Ramanna; this was possible by only a person of his calibre, his understanding and most importantly with his level of high-quality leadership. Raja Ramanna really moved India's nuclear programme with high momentum after the unexpected passing away of Homi Bhabha.

Years later when I took over as Director of National Institute of Advanced Studies (NIAS), the presence of Ramanna spending time at NIAS was an extraordinary opportunity for me to get his view of this unique institution in the context of J. R. D. Tata's vision. The vast knowledge of Ramanna, his interest in physics, India's nuclear programme, the need to develop science through aggressive research and above all his interest in other areas such as Western classical music truly reflected the multidisciplinarity of his interest. This in turn helped me to shape my own thoughts as a follow up of Prof. Rodham Narasimha's leadership. I often consulted Ramanna on a variety of matters – science, social sciences, mathematics as well as issues of National Security. We used to often meet at the cafeteria of NIAS where he and I were the last of the people to take our lunch. Under the quiet atmosphere of the cafeteria, we had long conversations on a variety of subjects which could be considered within the framework of NIAS's activities. He was truly a **Pitamaha** (Grandfather) and I will always remember the advantages I gained in providing my own leadership to NIAS with his vision and guidance.

Acknowledgement

I would like to appreciate Profs. V. S. Ramamurthy and M. B. Rajani for prompting me to write this contribution as a part of a volume being brought out by the National Institute of Advanced Studies to commemorate the birth centenary celebrations of Dr. Raja Ramanna. I would like to thank Ms. Brinda Nagarajan, my Executive Assistant at Raman Research Institute, for helping me in the preparation of this manuscript.

Dr. Krishnaswamy Kasturirangan is an Indian space scientist who headed the (ISRO) from 1994 to 2003 as its Chairman. Over many decades of a highly distinguished career, Dr. Kasturirangan has served in capacities advising governments on a range of policy interventions. He is presently Chancellor of Central University of Rajasthan and NIIT University, and Head of the Committee to formulate National Education Policy 2020. He is the former Chancellor of Jawaharlal Nehru University and the Chair of Karnataka Knowledge Commission. He is a former member of the Rajya Sabha (2003–2009) and a former member of Planning Commission of India (now reorganised as NITI Aayog). He was the Director of the National Institute of Advanced Studies, Bangalore, from April 2004 to 2009. He is a recipient of the three major civilian awards from the Government of India: Padma Shri (1982), Padma Bhushan (1992) and Padma Vibhushan (2000), and several national and international awards.

[Prof. K. Kasturirangan (24 October 1940 – 25 April 2025), who had strongly supported our efforts at Dr Raja Ramanna Centenary pro-grammes and bringing out this volume, passed away before it could be published. He provided his contribution despite his very frail health. We shall remain grateful to him for his guidance in our endeavours. (Ed.)]

Chapter 14

Dr. Raja Ramanna in the Defence Environment: Some Glimpses

K. G. Narayanan

Aeronautical Development Establishment, Bengaluru

Defence Research and Development Organisation, Bengaluru

kgnarayanan2@gmail.com

This is a small tribute to a very great person – Dr. Raja Ramanna, the scientist and philosopher, reminiscing mostly of my personal contacts with him in the defence environment. I cannot claim to have been a close professional associate or student of Dr. Raja Ramanna. I did not know much about him, until his name burst into national scientific consciousness as one of the key movers of the Peaceful Nuclear Explosion (PNE), from concept to creation. By May 1974, I had already been in the Defence Research and Development Organisation (DRDO) for over nine years. However, a nuclear explosion project of the Atomic Energy Department which could not be done without major contributions from DRDO had completely escaped my reckoning. Evidently, information secrecy was so effective that I was taken totally by surprise when my brother, an IAS officer, called me on the telephone from Gujarat on 18 May 1974, to convey his congratulations to DRDO scientists on the Buddha Smiled event!

Power to his nuclear elbow: It was a huge moment for the scientists in India, yet, I did not know much about the people who made it happen.

Then came an article on Dr. Raja Ramanna, the principal architect of PNE, in the *Illustrated Weekly of India* – decorated with a fabulous cartoon by RK Laxman. It portrayed a moustachioed, bespectacled scientist in his laboratory with smoke curling up from his cigarette. I remember that the write-up ended with the ardent wish of the cartoonist with the words "More power to Ramanna's elbow". That stayed in my mind and kindled my interest greatly. I took the opportunity to learn more about that phrase as well as Dr. Ramanna through chats with many DRDO friends who collaborated with him in the PNE – especially Shri N. S. Venkatesan and Shri M. Balakrishnan, both former Directors of the Terminal Ballistics Research Laboratory, Chandigarh. His kindness and graciousness in dealing with colleagues is remembered by many. In a recent retro conversation, Shri Balakrishnan cited several instances of the joy of working with Dr. Ramanna including how Ramanna had personally arranged for hot meals to be served to Balakrishnan and his team emerging from the pit at 2 AM after attending to last-minute technical issues.

Battle Tanks, Missiles and Fighters: When Dr. Ramanna arrived in DRDO as its Chief in 1978, his reputation preceded him (Fig. 1). He was known to be a scientific leader who made decisions quickly. He was also known to be wary of long file notings and his attention span during presentations was empirically estimated to be about 10 minutes. What could not be conveyed to him in a brief presentation or in a one-page note was not ready for decision at his level.

Fig. 1.   (From left to right) Dr. V. S. Arunachalam, Prof. D. S. Kothari, Dr. Raja Ramanna, Prof. S. Bhagavantam and Dr. B. D. Nagchaudhuri.

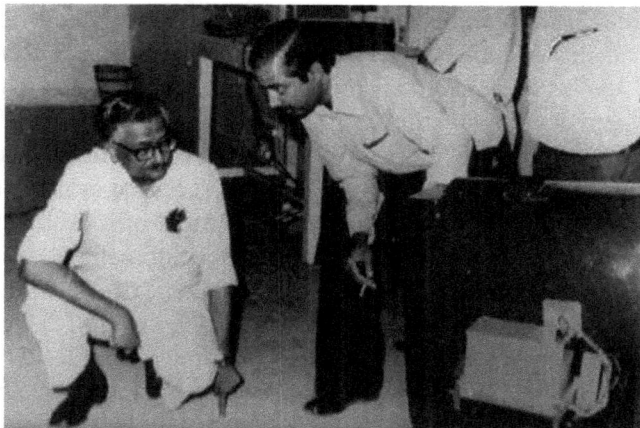

Fig. 2. Author (Right) briefing Dr. Ramanna (Left) at ADE, Bangalore.

So it was with a pleasant anticipation that my colleagues and I looked forward to his first visit to Aeronautical Development Establishment (ADE), hoping to convince the new boss that the R&D project proposals which we had submitted to headquarters earlier in the areas of flight simulation, remotely piloted vehicles and airborne sensors were indeed necessary and important (Fig. 2). We had prepared thoroughly to answer the numerous questions which had already been raised in long pending files by the earnest gatekeepers at New Delhi. We hoped that the new Chief could be convinced by us adequately to approve these projects on his return to DRDO Headquarters. What happened was quite different, thinking back, quite characteristic of Dr. Ramanna.

After a brisk walk through the laboratories and friendly chats with several scientists, he sat down to listen to the presentations. At the end of some 20 minutes, by which time my colleagues and I had merely listed the scope of our presentations, Dr. Ramanna put his hand up and said that he had understood what we were trying to do and that was enough for him. He pulled out his pen from his pocket and enquired "where do I sign". That was a real surprise and an embarrassment too because the official file was in New Delhi waiting to be submitted to him through proper channel. The situation was retrieved by getting him to sign on a mock file before he departed from ADE that day. It was evident that he was well ahead of the bureaucracy that surrounded him, and he believed in the dictum that *form should follow function.*

Dr. Ramanna's 4-year tenure (1978–1982) as the Scientific Adviser (SA) to Raksha Mantri (RM, Defence Minister) and DRDO Chief resulted in several significant reformatory moves with respect to the major weapon system development projects of DRDO at that time (Fig. 3). Main Battle Tank (MBT) at Combat Vehicles Research and Development Establishment (CVRDE), Avadi, Chennai, received some important course corrections with respect to the procurement of diesel engines and the development of targeting systems. The development of guided missiles received a major fillip with the establishment of a technical group consisting of experts from the Army, Navy, Air Force and DRDO to identify the tactical and strategic requirements of guided missiles to be developed. This commendable effort, backed by the induction of Shri Abdul Kalam into DRDO in early 1982, resulted in the formulation of the Integrated Guided Missiles Development Programme (IGMDP) a little after Dr. Ramanna handed over charge as SA to RM to his longtime mentee Dr. V. S. Arunachalam. IGMDP received the powerful support of Defence Minister Shri R. Venkataraman (Fig. 4).

Among the many structural and administrative reforms introduced in DRDO during his tenure was the Flexible Complementing scheme for the assessment and promotion of scientists based on the merit of their performance, without the constraint of sanctioned vacancies available. It was also Dr. Ramanna's love for conserving the

Fig. 3.   Raksha Mantri Shri R. Venkataraman (second from left) with Dr. Ramanna (extreme left) at Metcalfe House.

Fig. 4. Stars in conjunction: (left to right) Prof. D. S. Kothari, Prof. S. Bhagawantam, Dr. Raja Ramanna and Dr. V. S. Arunachalam.

heritage which led to the refurbishing of Metcalfe House a British Era building with a lot of history associated with it.

Perhaps Dr. Ramanna's role in bringing together diverse ideas and actors on the Indigenous fighter aircraft development is not so well known. Prof. Roddam Narasimha, the celebrated fluid dynamicist and a doyen of the Indian aeronautical engineering community, records a significant event in the genesis of Light Combat Aircraft (LCA) as follows:

So, the project (of identifying the characteristics of the indigenous fighter to be developed) continued with more vigour and we had discussions with IAF and DRDO ... ... In perhaps the first such meeting in Delhi, which took place in South Block, as I came out and was walking in the corridor, I saw Dr. Raja Ramanna, who had just assumed charge as Scientific Advisor to the Defence Minister. I had worked with him earlier in one or two committees on non-aeronautical matters, and the first thing he said was 'What are you doing here?' I said I had come for an LCA meeting. He said 'What is the LCA? I haven't heard of it. Can you come to my office after your meeting and tell me about it?'—which I did. Some months later Dr. Ramanna told me that he had discussed the LCA with the Air Force and DRDO; they were interested in the concept but were not sure about the performance estimates (projected by us). He told me. ... that our ideas and figures should be tested in presentations to some of the major European industries, and he would like me to lead a small delegation with members from Indian Air Force (IAF), DRDO, Hindustan Aeronautics Limited (HAL) and National Aerospace Laboratories (NAL) (to Europe). ... ... Some five or six of us visited Messerschmitt-Bölkow-Blohm (MBB) at Munich, Dassault at Paris, British Aerospace at London, and Svenska Aeroplan AktieBolaget (SAAB) at Stockholm.

Finally, an agreement was reached among the members of this group (after the interactions with the European manufacturers). It was based on this congruence of ideas of IAF, HAL, NAL and DRDO that Dr. Ramanna, as Secretary of Department of Defence R&D, submitted the proposal for the development of the Indigenous fighter (LCA) to the Cabinet for approval. The approval for this critical seed project came through some months after Ramanna left DRDO to return to DAE.

Run Silent, Run Deep: Another momentous decision of the Government of India taken around 1988 was to design and build an India-designed nuclear-powered submarine to boost the strategic capabilities of the Indian Navy, largely persuaded by Dr. Raja Ramanna and his long-time friend in the Indian Navy, Vice Admiral MK Roy. The lead design role and construction responsibilities were assigned to a special group in the Navy. The mini-nuclear reactor propulsion system was to be developed by BARC. Several equipment, sonars and other sensors and special steels and other materials were to be developed by DRDO laboratories. The financial provisions for this Advanced Technology Vehicle (ATV) were made in the annual budgets of the Department of Defence R&D; the SA to RM chaired the apex committee to monitor progress. So, Dr. Ramanna set the tone for this unusual and highly strategic project to proceed smoothly across the walls of Departments, ably followed by his successor Dr. V. S. Arunachalam and others. Dr. Ramanna would indeed have been delighted to see the historic event of INS Arihant becoming operational, as a nuclear-powered, strategic missile launch capable, system of the Indian Navy in 2016.

Trouble in the Skies: Some years later, I had yet another opportunity to interact with Dr. Ramanna personally when he was appointed Raksha Rajya Mantri (Minister of State for Defence) – perhaps the only occasion when an eminent scientist was inducted into the Union Council of

Ministers. The year was 1990 and I was the Director of ADE which had taken up the responsibility of developing the Digital Flight Control System (DFCS) for the LCA. My boss, Dr. V. S. Arunachalam, asked me to brief Dr. Raja Ramanna on the technical features of what was popularly known as the Fly-By-Wire flight control.

The context was the grim situation in the Indian aviation sector after the Government of India ordered grounding of all the 14 Airbus – 320 aircraft in the inventory of (then) Indian Airlines (IA), after one such IA aircraft crashed in Bangalore killing 92 persons, in February 1990. The Government of India had even let it be known that they would lease out the grounded aeroplanes to other operators and sell off the four aircraft yet to be delivered by Airbus. There was confusion in the minds of people on what exactly led to this fatal crash. As the Airbus 320 was the very first commercial aircraft in the world to incorporate a fully computer-controlled flight control system (called Fly-By-Wire), doubts arose about its performance and whether the pilots were adequately adapted to the nuances of this technology. Political factors also entered the scene in a big way. IA continued to incur heavy losses by the continued grounding of their brand-new Airbus 320 fleet for several months – despite a committee of technical experts headed by an Air Marshal of IAF as well as the Directorate General of Civil Aviation (DGCA) clearing the aircraft for further passenger service. The situation was most confusing, to say the least. As Dr. Raja Ramanna was the Scientific Adviser to Defence Minister who at that time was also the Prime Minister, his advice was sought on the technical aspects of the issue.

I headed to Dr. Ramanna's residence at RT Nagar in Bangalore one afternoon accompanied by my colleague, Rear Admiral S. K. Ray who oversaw the LCA fly-by-wire flight control system development. Contrary to my earlier experience, Dr. Ramanna heard our presentation patiently. We explained how triple or quadruple redundancy schemes in the design and engineering aspects of the Fly-By-Wire system reduced the probability of loss of control so much as to be negligible. In the subsequent discussions, he confined himself mainly to issues related to what is technically known as the Pilot–Vehicle Interface. That struck me as very astute, considering that it was clearly the area where the cause of this accident (and some subsequent accidents) was found eventually. He also questioned us keenly on the procedure adopted in India for certification of airworthiness and quality assurance of commercial aircraft manufactured elsewhere and the skill sets of personnel issuing such clearances in India.

While his depth of knowledge in this area surprised me at first, I reminded myself that Dr. Raja Ramanna had actually walked out to safety from the burning wreck of a fatal Air India Boeing crash in Bombay in 1982! I believe, Dr. Ramanna played an important role in the Government rescinding its irrational order of grounding the Indian Airlines fleet of Airbus 320.

Philosophical and Computational: Another sombre event which comes to my mind is a presentation made by a DRDO group working on Operations Research and Systems Analysis to Dr. Ramanna as Raksha Rajya Mantri. Dr. Arunachalam had invited me also to attend the meeting. The subject had to do with scenario modelling and simulation studies on nuclear weapon strike exchange in the Indian context. Dr. Ramanna listened most attentively, made interventions and unusually made notes. When the meeting ended, he walked out very silently without the usual handshakes. He invited Dr. V. S. Arunachalam and me to his office for some coffee on that bitterly cold evening. It took him a while to relax in his normal easy manner. He was philosophical and quoted verses from Mahabharata.

My interactions with him continued when Dr. Ramanna became the Director of National Institute of Advanced Studies, Bangalore. I was surprised when I received a telephone call from him when he asked me if I was using MATLAB software (a new computational tool used in Physics and Engineering) and whether I could help him learn and use it. I admitted to him, somewhat in shame, that my knowledge of MATLAB was sketchy. I could request one of my colleagues who was using this tool extensively to go over and assist him. That is how it was done, leaving my young colleague and myself wondering about the mental capacity and interests of Dr. Ramanna.

Easy-going Superman: How does one remember him? A visionary, a scientist of great integrity and achievements, and a musician of repute. Besides all that, he was also a friendly man who was easy to talk to on many subjects – science, arts, philosophy, evils of prohibition or the best way to make a rum cocktail. Dr. Rukmini Sankaran, former Director of Defence Food Research Lab at Mysore, sums up the achievements of her lab as follows: "I have one regret though. Dr. Raja Ramanna, the great visionary and the then SA to Raksha Mantri, had often asked us to mimic the flavour of 'Avarekkalu' (a type of bean seed seen typically in

Karnataka) into many others (ready foods for the soldiers). Flavour is such a complex issue that, however much we tried, I had to admit apologetically that we could not succeed". Quite often, Dr. Ramanna shot questions at me in Kannada and smiled indulgently when I struggled to answer.

Dr. K. G. Narayanan completed his career in R&D as the Chief Advisor at DRDO Headquarters after having been a member of its scientific workforce from 1965 to 2002, contributing to several of the development programmes in solid state device technology and aeronautical engineering. He led two of the major laboratories: Aeronautical Development Establishment (ADE) and Defence Avionics Research Establishment (DARE), Bangalore, from 1986 to 2001. He was also the first Director of the privately funded SSN Research Centre, near Chennai. Recently, he co-authored the book *Digital Flight Control Systems for Practising Engineers* and edited a historical account of DRDO's work titled *Endeavors in Self Reliance*. Both the books were published by DRDO. He is a Fellow of the Indian National Academy of Engineers, a Distinguished Alumnus Award winner of IIT, Kharagpur, and a recipient of several professional awards including the Systems Society Gold Medal and the D. S. Kothari Award of the Indian Science Congress Association.

Chapter 15

Scientific Contributions of Dr. Ramanna

V. K. Saraswat

NITI Aayog

Defence Research and Development Organisation

Indian Minister of Defence, New Delhi

vk.saraswat@gov.in

Dr. Raja Ramanna was a multifaceted personality – an eminent nuclear physicist, a highly accomplished technologist, an able administrator, an inspiring leader, a gifted musician, a scholar of Sanskrit literature and philosophy and above all a completed human being. He made important theoretical and experimental contributions in various areas of nuclear physics.

Dr. Ramanna is regarded as one of the most successful creators of Science and Technology in India. He was not a so-called ivory tower scientist. Following the ideals of his illustrious predecessors Homi Bhabha and Vikram Sarabhai in India's nuclear energy programme, Ramanna played an important role in placing the country's indigenous nuclear capabilities on a firm footing and in this process his contributions towards shaping India's energy and security programmes are quite significant.

In fact, Dr. Ramanna's contribution to India's peaceful nuclear explosion experiment is well known. India's first peaceful nuclear experiment was carried out underground in the Rajasthan desert on 18 May 1974, which asserted the technological advancement India had determined to perfect in the post-independence era.

Dr. Ramanna's Life: Dr. Raja Ramanna was born in Tumkur in Karnataka on 28 January 1925. He had his early education in Mysore and Bangalore. When his family shifted to Bangalore, Ramanna joined the Bishop Cotton School. From Bishop Cotton School, he went to St. Joseph's School for his intermediate studies.

After completing his intermediate studies at St. Joseph's, Bangalore, he joined the Madras Christian College in Tambaram. He was among the six students who were selected for B.Sc. (Honours) course majoring in physics. After obtaining his B.Sc. (Honours) degree, he went to England to work for his doctoral degree in the field of nuclear physics at King's College, London, as Tata Scholar.

During one of his trips to London in 1947, Dr. Homi J Bhabha offered Ramanna a job at the Tata Institute of Fundamental Research, the cradle of India's atomic energy programme. Bhabha allowed Ramanna to complete his Ph.D (1948).

Ramanna joined the TIFR on 1 December 1949. Bhabha, who had known Ramanna's interests and abilities in music, allotted him two adjacent rooms on the top-most fourth floor of the hostel: one for Ramanna and the other for his piano. The ground floor of the hostel became the nuclear physics laboratory of Ramanna, where he started his work on nuclear fission and scattering.

Ramanna made important contributions in several areas of neutron, nuclear and reactor physics. Ramanna played a leading role in organising physics and rector physics programmes at the Bhabha Atomic Research Centre, Trombay. Ramanna was a young reactor physicist in the team under Bhabha, when India's first research reactor, Apsara, was commissioned on 4 August 1956.

As a part of the studies relating to the design and construction of **Apsara, India's first reactor**, Ramanna studied the process of neutron thermalisation in several moderating assemblies. Ramanna and his group determined the neutron diffusion and slowing down constants in water and beryllium oxide by using a pulsed neutron source. The neutron spectra emerging out of these moderating assemblies were also studied.

Apsara, once commissioned, made intense thermal neutron beams available for basic research. This prompted Ramanna to undertake a programme of experimental investigations of secondary radiations emitted in thermal neutron-induced fission of U235.

Ramanna and his co-workers measured the energy and angular distributions of prompt neutrons and gamma rays emitted by fission fragments. Such measurements provided important information on the times of these radiations, the presence of scission neutrons, the average spin of the fission fragments and so on.

The stochastic theory of fragment mass and charge distributions in fission is a unique contribution of Ramanna to fission theory. Geometrical interpretation of atomic and nuclear binding energies was another novel contribution of Ramanna and his group.

Ramanna was a staunch patriot. He could have easily settled abroad but he spurned the charm of living in a developed country and responded to the call of Homi Bhabha and joined India's effort to develop a strong indigenous base of science and technology. He helped create an efficient manpower in the country.

Raja Ramanna was an Able Administrator

Dr. Raja Ramanna held numerous high-level positions throughout his career, including the following:

Leadership roles in prestigious institutions such as the Bhabha Atomic Research Centre (BARC), Defence Research and Development Organisation (DRDO) and Atomic Energy Commission.

Government positions as Scientific Advisor to the Minister of Defence and Minister of State for Defence.

Founding and directing key scientific institutes such as the National Institute of Advanced Studies, Bengaluru, and the (now Raja Ramanna) Centre for Advanced Technology, Indore.

Dr. Ramanna was instrumental in building India's nuclear programme and contributing significantly to its research and development. He also made substantial contributions to other scientific fields and was a strong advocate for scientific progress. His passion for his work and his dedication to nation-building are evident in his numerous achievements and accolades.

Dr. Ramanna's Contribution to Defence Research

Though at the beginning of his tenure as Scientific Adviser in 1978, Dr. Ramanna, who was still Director of BARC, commuted between Mumbai and Delhi regularly, but soon he became a fulltime Scientific Adviser. He made it a point to visit all the laboratories and field stations of the DRDO more than once and this gave him the opportunity to explore different manifestations of nature in places such as Kumaon, Lahaul, Pygmalion Point and Andaman and Nicobar Islands and feel rejuvenated.

In this process, he had the time and opportunity to appreciate the good work being done by the small laboratories and the field stations at Manali, Almora, Jodhpur and other places. The scientists of these institutions were encouraged by the visits and interaction with the Scientific Adviser and showed it with an upward trend in the quality and quantity of their output. His passionate desire to preserve the nation's historical heritage found expression in the renovation and maintenance of the historic building in Delhi, the Metcalfe House.

Dr. Ramanna was not a stranger to DRDO. His first acquaintance with the Organisation was around 1957 when he attended the Defence Science Conference convened by Dr. D. S. Kothari the Scientific Adviser at that time. He had close interaction with Dr. B. D. Nagchaudhuri and Mr. N. S. Venkatesan, who was Director of Terminal Ballistics Research Laboratory (TBRL), in the pre-1974 days when DRDO collaborated with the Department of Atomic Energy for the peaceful nuclear explosion. From these experiences, he was well aware of the essential cultural dissimilarities between the DRDO and the BARC/DAE.

Immediately after assuming the office of the Scientific Adviser to Raksha Mantri (SA to RM) Dr. Ramanna swung into action and in about a fortnight, issued a letter to all the laboratories and field stations of DRDO about the restructuring of DRDO Headquarters and distribution of work among the Chief Controllers, Research and Development (CCR&Ds) to reduce the bureaucratic overheads and ensure faster response to the needs of the laboratories.

On technical activities, the link between the Heads of the Laboratories and the Director General R&D would be direct. The Heads of Laboratories would keep their respective CCR&Ds in the picture so that the latter would be able to effectively interface with the Services. The CCR&Ds would be the single point contacts for their laboratories on all issues requiring clarification/action from multiple directorates at DRDO Headquarters. The role of Headquarters personnel would be to assist the laboratories in all possible ways by providing scientific and administrative support and data not available to the laboratory personnel. In addition, they would keep the Scientific Adviser informed of all the modern developments in their respective areas and effectively liaise with the Services and concerned Ministries of the Government.

New Personnel Policy at DRDO

Dr. Ramanna brought about changes in the structure for recruitment and assessment of staff in DRDO. Until then, the recruitments had been through the Union Public Service Commission (UPSC), which was not the best route to spot scientific and engineering talents. The promotions were vacancy-based and not totally merit-dependent and there were many budgetary constraints that were stringent and unreasonable.

Dr. Ramanna's predecessor Dr. M. G. K. Menon had devoted some effort to change the structure, and following, this he succeeded in convincing the UPSC to introduce flexible complementing that enabled *in situ* promotions for competent scientists and engineers (it would require a few more years and some more efforts before DRDO could become independent of UPSC for recruitment and assessment).

There were apprehensions in the minds of many scientists at DRDO Headquarters and at the Laboratories about the impact of the new personnel policy on their careers, the methodology of implementation and the delays that might arise due to legal challenges. Speed was of essence and here again, Dr. Ramanna showed his penchant for action to reduce the agony of uncertainty. Dr. Raja Ramanna firmly put in place the new personnel policy for which the scientists of DRDO owe them a debt of gratitude. The scientists of DRDO could all look forward to merit-based promotions, no vacancy-based constraints, regular and annual holding of assessment boards and all promotions given effect from a specified date.

Department of Defence Research and Development

Even though the Scientific Adviser enjoyed the status of the Secretary to the Government of India, DRDO was put under the Department of Defence Production from the days of Dr. Nagchaudhuri. Dr. Raja Ramanna took the step of creating the Department of Defence Research and Development so that the important papers requiring approval/concurrence from the Cabinet, the Prime Minister, Raksha Mantri or Raksha Utpadan Mantri or other Ministers could be sent straight to the destination. In all cases of development of major weapon systems, direct communication with the political head of the Government would be necessary because it involved strategic decision-making.

The creation of a separate department helped in reducing the paper-work on important issues and the corresponding delays for DRDO. The enormous prestige Dr. Ramanna enjoyed with the political establishment of the country as the prime architect of the peaceful nuclear explosion stilled any action to stall the creation of a separate department for research and development. The main areas of responsibility for the department were rendering advice to the Defence Minister and the three Services on all scientific aspects of military operations, equipment and logistics, formulation of research and development plans, administration of Defence Research and Development Services (DRDS) Rules, framing of personnel policies and providing the backup for the Defence Research and Development Council.

Streamlining the Operations within

Dr. Ramanna analysed that the laboratories of DRDO in major disciplines had reached a stage where the Organisation was confident of taking on the development of major systems for the Services. He found that the cost of Navy's projects was a distant third to the other two Services as the Navy did not have any major weapon system or sensor system projects at the laboratories dedicated exclusively to meet their needs. He decided to increase the budget for the Naval Science and Technological Laboratory (NSTL) so that the laboratory could accelerate the process of building infrastructure and test facilities for the development of naval weapon systems.

He subsequently focused on laboratories within the Armament and Materials and General Stores divisions, which housed the most numerous smaller projects. Following discussions with Mr. N. S. Venkatesan, Director of Armament Research and Development Establishment (ARDE), and his top researchers, ARDE committed to shifting from a reactive approach to a proactive strategy, aiming to reduce the number of projects while increasing their scale in the future.

In the Materials Group, Dr. Ramanna found that Defence Metallurgical Research Laboratory (DMRL) was already very active and interacting with other laboratories with respect to their needs and had very good interaction with the production agencies. With the other laboratories in the Materials Group, he urged them to drop smaller projects, which could be handled by institutions other than DRDO and try to focus on bigger issues.

In the case of the Aeronautics Group, he got Aeronautical Development Establishment (ADE) to switch over to work on a major programme of developing Pilotless Target Aircraft (PTA) for meeting the Services requirements.

For Defence Research and Development Laboratory (DRDL), Dr. Ramanna was on the lookout for a scientist who had a good track record and who was dedicated and committed to his work and profession. He found his man in Dr. A. P. J. Abdul Kalam who was at the Department of Space and persuaded him to take up the challenge of building the missiles for the country. This resulted in the successful development of missiles of different ranges (Prithvi, Nag, Akash, Agni, etc.).

When the Indian Air Force indicated that they would need replacements in the 1990s for the ageing fighter fleet, Dr. Raja Ramanna in his capacity as the Scientific Adviser took the lead in February 1980, forming an internal group of scientists to conduct a detailed study of the various aspects of the Light Combat Aircraft (LCA) and related engine development programmes. The study would also include the assessment of the resources and technologies from within the country.

Preliminary design configuration studies based on a number of available engines including the indigenous GTX of Gas Turbine Research Establishment (GTRE) were completed. The concept as envisaged by the committee of the future fighter aircraft and the new advancements likely to find a place in the proposed solution to the Air Force's needs were enumerated.

Fostering Inter-Lab Interaction

Dr. Ramanna's own experience of the peaceful nuclear explosion had brought home to him the interdisciplinary nature of modern weapon systems. Therefore, he set about improving the collaboration among the laboratories. He set up Advisory Committees for each group of laboratories with Chief Controllers in the chair to bring interdisciplinary perspective to the activities of each laboratory. He discouraged nodal systems laboratories.

Dr. Ramanna fostered the concept of laboratory complexes in Bangalore, Delhi, Hyderabad and Pune so that optimum use of the facilities and sharing of expertise would be possible among the laboratories. Therefore, he stopped the permanent building project of Electronics and Radar Development Establishment (LRDE) at High Grounds in Bangalore. Instead, he wanted the permanent location of the laboratory to be shifted to the area where the other two laboratories, namely ADE and GTRE were situated (presently called CV Raman Nagar). With this in view, he obtained additional land from the Karnataka Government adjacent to GTRE and permitted LRDE to have its laboratories situated on that site. On the land contiguous to LRDE, he allowed the present residential complex for all DRDO institutions of Bangalore to be constructed.

Impact on DRDO

Dr. Raja Ramanna brought home three important aspects of leadership. **First**, the analytical power of the intellect as a tool for timely and decisive action. **Second**, it is important to keep channels of communication open with the decision-makers at the user end and with those in the political establishment. **Third**, team building across the Organisation is necessary and nodal laboratories in major systems development must reach across to other laboratories in the organisation.

When it was known that he would be shortly moving out of DRDO because of his being appointed as Chairman of Atomic Energy Commission, Dr. Raja Ramanna persuaded the Government not to look outside the Organisation for his successor. He considered that there was more than one senior scientist within the DRDO who could head the Organisation. Dr. V. S. Arunachalam became the first DRDO scientist to head the DRDO and also to assume the office of the Scientific Adviser.

Personal Experience with Dr. Raja Ramanna

1. **Project Valiant:** Dr. Ramanna witnessed the test be action and waited for the technical team to execute the launch from the viewpoint (waited 4 hrs for a technical problem to be resolved – *"Let the scientists solve the problem"*). He also had an extended personal interaction with the scientist after the launch during the lunch session.

2. **Under Project Valiant IRBM:** With a 1000 km range, project outline was greatly appreciated by Dr. Ramanna, termed as the most feasible project. By the end of 1981, the engine was ready. However, Dr. Ramanna's departure from DRDO put the project on the back burner, which would have led to the early development of intermediate-range ballistic missile (IRBM) in India.

3. **Prithvi Missile:** Dr. Ramanna was critically involved in the Prithvi Project. Prithvi Project was recognised as the potential project that would *see light of the day at the earliest.*

Conclusion

Upon retiring as Head of the Atomic Energy Commission in 1988, Dr. Ramanna was invited by J. R. D. Tata to establish a new research institute in Bangalore. Tata envisioned this institution to serve a dual purpose. First, it would offer comprehensive learning programs for mid-level professionals from both the public and private sectors. These programmes aimed to expand their knowledge base and enhance decision-making abilities, thereby influencing the nation's progress in science, technology, business and governance. Second, the institute would undertake in-depth research across various disciplines, with a particular focus on social science projects that could positively impact society. This institute is known as the National Institute of Advanced Studies.

Dr. Ramanna's life was a testament to the belief that humanity's purpose lies in serving others. His engagement extended far beyond professional confines, encompassing civic duties, national policy and a profound interest in the complexities of Indian society. This breadth of involvement cultivated a compassionate and empathetic worldview.

A beacon of optimism and perseverance, Ramanna's life inspires individuals to become active participants in shaping a better world. His unwavering dedication to addressing societal challenges, coupled with his ability to find solutions amidst adversity, provides a compelling model for leadership.

Ramanna had a deep interest in music. He himself was an accomplished musician. He wrote a book on music, *The Structure of Music in Raga and Western Systems.* He was actively involved in setting up the Bangalore School of Music. Ramanna had interest in philosophy. He also took keen interest in yoga. He had a sense of humour that was subtle and enjoyable. He was a very simple person, and he was approachable to all.

Transitioning seamlessly from high-level government positions to roles focused on broader societal issues, Ramanna demonstrated exceptional adaptability. His ability to connect with people from all walks of life, from spiritual leaders to scientists, was underpinned by a sharp wit, keen intellect and unwavering respect for others. Despite his influential positions and numerous achievements, he remained remarkably grounded, exhibiting equanimity and detachment.

Ramanna's intellectual pursuits were as diverse as his professional endeavours. He excelled in fields as disparate as science and spirituality, leaving an enduring legacy in each. A fervent admirer of India's ancient wisdom, he was a deep thinker who synthesised philosophical concepts from Advaita and Buddhism. His unique perspective and ability to connect disparate ideas resulted in original and insightful contributions to various fields.

Ultimately, Dr. Ramanna's life was characterised by a rich tapestry of experiences and accomplishments. His enduring impact lies in his unwavering commitment to human welfare, his ability to inspire others and his intellectual depth.

Vijay Kumar Saraswat is a well-known Indian scientist who formerly served as the Director General of the Defence Research and Development Organisation (DRDO) and the Chief Scientific Advisor to the Indian Minister of Defence. Dr. Saraswat is presently a member of NITI Aayog, the Indian Government's apex public policy think-tank and the President of Sree Chitra Tirunal Institute for Medical Sciences and Technology, Trivandrum. He is also a former Chancellor of Jawaharlal Nehru University. Dr. Saraswat is the key scientist in the development of the Prithvi missile and its induction into the Indian armed forces. He is a recipient of the Padma Shri and Padma Bhushan from the Government of India, and a fellow of National Academy of Engineering, Aeronautical Society of India, Astronautical Society of India and Institution of Engineers.

Chapter 16

From Student to Scientist: Dr. Ramanna's Enduring Influence

K. Siddappa

Microtron Centre, Mangalore University
and
Bangalore University, Bengaluru, India

siddappakadajji@gmail.com

I consider myself immensely fortunate and privileged to have known Dr. Raja Ramanna, India's foremost nuclear scientist and a remarkable patriot. It brings me great joy to share some memorable moments from my long association with him on the occasion of his centenary celebration.

I first met Dr. Raja Ramanna in 1968, during a special lecture he delivered in the Department of Physics at Manasagangotri, University of Mysore. At the time, I was in my final year of an M.Sc. in Physics, specialising in Nuclear Physics. Dr. Ramanna spoke on "India's Nuclear Energy Programme". It was a captivating lecture, delivered in his unique style, blending depth with wit and humour. Starting with the basics of nuclear fission, he skilfully explained the construction of nuclear reactors and their various peaceful applications. At the end of his talk, he patiently answered all our questions, leaving us inspired and motivated. His lecture was pivotal for me – it sparked my passion for research in nuclear physics.

After completing my M.Sc., I joined the Nuclear Physics Department at Andhra University as a CSIR research fellow in 1969. The same year, I had a chance to attend the Department of Atomic Energy Symposium on

Nuclear Physics and Solid-State Physics at Roorkee, where I was thrilled to meet Dr. Ramanna again and to hear his inspiring inaugural address. His talk offered a broader perspective on research across multiple disciplines and revealed the depth of his personality and intellect.

I introduced myself to him, mentioning his lecture at Mysore University the previous year and how it had motivated me to pursue research in nuclear physics. He was pleased to hear this and encouraged me warmly. That day, I also had the privilege of attending a talk by one of Dr. Ramanna's distinguished students, Dr. V. S. Ramamurthy, on the finer points of nuclear fission, which sparked a lot of positive discussions. Both Dr. Ramamurthy and Dr. Ramanna engaged with the audience, answering questions and clarifying doubts. My brief conversation with Dr. Ramamurthy marked the beginning of a friendship that would grow over the years.

In 1971, I again attended the Symposium on Nuclear Physics and Solid-State Physics, where I presented my first research paper. Dr. Raja Ramanna, who inaugurated the symposium, was present and asked me a question about interpreting my experimental results using the optical model. Though I'm not certain I answered his question correctly, I felt both excited and honoured to engage with him. That same evening, Prof. E. C. G. Sudarshan gave a special lecture, making it a memorable day.

The title of Prof. Sudarshan's talk was very intriguing, but the content was challenging to grasp. After the talk, Dr. Raja Ramanna, who chaired the session, began his presidential remarks in his characteristic style: "Well friends, judging by your puzzled faces and complete silence, I gather that most of you, like me, struggled to follow Dr. Sudarshan's talk, with its mathematical derivations and complex theories. But don't worry – relax, I'm also in your group".

To everyone's relief, he proceeded to break down the main points, providing a simplified explanation of the theoretical derivations and the essence of Dr. Sudarshan's talk, including its importance. Within about 20 minutes, he'd conveyed the significance of the lecture, winning an enthusiastic round of applause from the audience. That was quintessentially Dr. Raja Ramanna, everyone felt.

All this left me astonished at Dr. Raja Ramanna. Here was a renowned nuclear physicist, leading one of the world's top nuclear fission groups at the Bhabha Atomic Research Centre (BARC), with an impressive background in theoretical physics and an unparalleled ability to explain and interpret complex topics with wit and humour. But beyond his professional brilliance, he was a humble and deeply humane person. His character inspired me in ways beyond words.

In September 1977, when the Variable Energy Cyclotron was commissioned, an international conference was held in Calcutta (now Kolkata). I attended this event as part of a group from Mysore University, with Prof. B. Sanjeevaiah as our group leader. By then, I was already a faculty member in the Physics Department at Mysore University, having joined in 1973 after completing my Ph.D. in Nuclear Physics.

One evening, the Director of BARC, Dr. Raja Ramanna, hosted a dinner party for the conference delegates. The six of us from Mysore University were seated together, discussing the day's sessions, when Dr. Ramanna made his way around the tables, greeting everyone and encouraging us to enjoy the evening. Spotting our group, he approached with a friendly smile and said, "Oh! Groupism here too – Mysore people sitting together and gossiping! Fine, let me join in and be part of the conspiracy". With that, he pulled up a chair and joined us.

As he chatted with us, he suddenly noticed the glass in my hand and decided to tease me. "Young man, what's that in your glass?" he asked. Startled by his sudden question, I answered slowly, "Sir, it's lemon juice". "What? Lemon juice?" he exclaimed with mock horror. "A great dishonour to the host, and an even greater dishonour to the conference! Look around, Siddappa – even the lady delegates are holding beer or wine glasses!" Laughing, he called over a waiter and ordered beer and wine glasses. The waiter promptly brought them, and Dr. Ramanna insisted, "Take either one you prefer – or both – but not that lemon juice! No fussing, okay!" He laughed heartily and then greeted the others, leaving us all amused.

This was yet another remarkable experience for me with Dr. Raja Ramanna. Here was a great scientist and Director of one of the world's largest and most renowned research laboratories, BARC, moving around the room to greet delegates and perform his hosting duties with such humility, warmth and affection. At the same time, he was fully immersed in the enjoyment of the evening, making sure everyone felt included. I felt that only Dr. Ramanna could create such an atmosphere. It was incredible that such a towering figure with countless accomplishments treated me, a young novice, as a friend. This spoke volumes about Dr. Ramanna's simplicity and grace.

In 1981, Dr. Raja Ramanna visited the newly established Mangalore University as Chairman of its Advisory Committee. Previously, it had been a postgraduate centre of Mysore University, where I was working as a lecturer in Physics. When it was upgraded to university status, however, it lacked the necessary physical and academic infrastructure. Despite the Vice Chancellor's repeated efforts to secure support, little progress was made.

Dr. Ramanna and his committee toured the departments, assessing the facilities and gathering feedback from faculty members. The following day, he convened a press conference, where he publicly criticized the government and relevant authorities in his unmistakable style: "It is a mockery to establish a university in a remote location like this without basic facilities like water, electricity, transportation, and even minimal academic infrastructure. The science laboratories here are worse than high school labs, yet the name boards proudly display 'Department of Postgraduate Studies and Research'! This is indeed a strange experience, and it is beyond comprehension".

The press covered his remarks extensively, and the government quickly took notice, finally responding with the necessary support to develop the university. Dr. Ramanna's influence was powerful in Karnataka, his home state, where his contributions to research and development were highly respected. His reputation, especially following the success of the "Smiling Buddha" project and India's first Pokhran nuclear test under his leadership, brought national and international recognition to the state and the country.

In 1988, Dr. Raja Ramanna visited Mangalore University on an important occasion, as the Department of Physics had active collaborations with Bhabha Atomic Research Centre, Mumbai, Indira Gandhi Centre for Advanced Research (IGCAR), Kalpakkam, and Raja Ramanna Centre for Advanced Technology (RRCAT), Indore. During this visit, he stayed for five days after inaugurating a Regional Workshop on Experimental Physics. On the third afternoon, Dr. Ramanna approached me and said, "Siddappa, let us go out in the evening. Take someone with you, one of your good friends". I promptly made the necessary arrangements for the outing.

After boarding the car, he explained, "We will go to a good restaurant where we can enjoy some good food. Ask your friend to join us". After a brief pause, he added, "One condition: today, I am going to pay the bill". Surprised, I hesitated and replied, "Sorry, Sir. Today it is my pleasure to put the bill, as I have resumed my duties as a professor".

He was visibly happy and congratulated me, adding, "I have observed that once someone becomes a professor, they often stop doing research, thinking that professorship is the ultimate position and there's no need to continue their research work. I wanted at least one university in Karnataka to become a prominent research centre in nuclear physics and to encourage professors to continue their research. Unfortunately, everyone has disappointed me. Now, Dr. Siddappa, since you have become a professor, I want you to continue your research and develop a strong research centre in Nuclear Physics. I hope you will not disappoint me". I assured him, "Sure, Sir, I will not disappoint you". This conversation left me in awe of his passion for nuclear physics research and his keen interest in developing robust research centres within the university system.

We continued discussing various relevant issues related to the university system until we arrived at the restaurant. Dr. Ramanna inquired about the apparent gap between Vice Chancellors and Professors, questioning why professors maintain a distance from Vice Chancellors, even though Vice Chancellors were also professors before their appointments. Upon reaching the restaurant, my friend Dr. R. Shankar, an alumnus of IIT Bombay, was waiting for us. I introduced him to Dr. Ramanna, and we all sat together.

When the waiter came to take our order, Dr. Ramanna suddenly asked, "Dr. Siddappa, are you still a teetotaller?" Puzzled, Dr. Shankar responded, "I didn't understand, Sir. Why are you asking this?" Dr. Ramanna explained, "Look, Shankar, Siddappa made quite a fuss in Calcutta a couple of years back during an international conference dinner party. I tried to introduce him to beer or wine at least, but he refused and took only a glass of lemon juice! I have not forgotten that". I was astonished by Dr. Ramanna's excellent memory.

As we ordered, Dr. Ramanna surprised us by requesting chilli bhajis, a treat he eagerly anticipated. "Chilli bhajis are very good for health", he said with a wink. "Green chillies help lower cholesterol – or at least, that's what I like to believe since I enjoy them so much". His humour and easy-going nature filled the room with laughter, making us all feel at ease.

Our discussion continued as Dr. Ramanna expressed his concerns about the poor functioning of universities in the country, attributing it to the indifference of Vice Chancellors and the apathetic attitudes of professors. He highlighted the resultant injustices to the students' learning processes and the declining quality of research. Reflecting on the evening,

I realized what a memorable day it had been, witnessing Dr. Raja Ramanna's unwavering commitment to improving the academic and research standards of Indian universities.

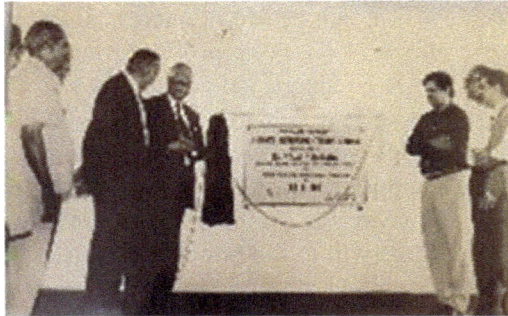

Fig. 1. Dr. Raja Ramanna (on the left of the plaque) inaugurating the Science Instrumentation Centre at Mangalore University, established by the author.

Dr. Raja Ramanna's presence on the day I became a professor and receiving his blessings marked a pivotal moment in my career. His encouragement inspired me, instilling a new level of confidence. This propelled me to complete my ongoing projects with renewed vigour and to take on ambitious new R&D challenges. The following six years of hard work yielded remarkable results. Starting from scratch as the Founding Director of the UGC-sponsored University Science Instrumentation Centre, I established a vital facility for the science departments with comprehensive infrastructure (Fig. 1). I launched an M.Sc. course in Electronics with industry collaboration and established a fully equipped Electronics Department as its Coordinator. More importantly strengthened environmental radioactivity research by equipping the laboratory with advanced measuring instruments and spectrometers with the help of a larger project from the Board of Research in Nuclear Sciences, Department of Atomic Energy (BRNS, DAE). One particularly challenging project during the period was establishing a variable energy electron accelerator – the Microtron – in collaboration with RRCAT and BARC, which we completed successfully.

In 1995, Dr. R. Chidambaram, then Chairman of the Atomic Energy Commission, commissioned the Microtron in the presence of Dr. Raja Ramanna. An international conference titled "R&D Using Electron

Accelerators" was organised on this occasion. The conference was attended by senior scientists from BARC, RRCAT and IGCAR, as well as researchers from universities, industry and foreign delegates. Dr. Ramanna inaugurated the event with an inspiring speech of course; he did not forget to suggest conducting nuclear physics research investigations and also using a microtron facility as it provides high-energy bremsstrahlung photons with suitable converters and thermal neutrons in addition to high-energy electrons. His talk set a positive tone for the deliberations. He attended all three days, actively participating in the sessions and offering insightful comments and suggestions (Figs. 2–5).

Fig. 2. Dr. Raja Ramanna (second from left), Dr. R. Chidambaram (third from left) and the author (extreme right) on the occasion of the Commissioning of Microtron.

Fig. 3. Dr. Raja Ramanna and Dr. M. I. Savadatti on the occasion of 15th Foundation Day Celebration of the University Science Instrumentation Centre.

Fig. 4.   Dr. Raja Ramanna Inaugurating the Electronics Department Building.

Fig. 5.   Dr. Raja Ramanna (centre in red tie) in a discussion with Dr. R. Chidambaram (on his left) and others.

Following the conference, Dr. Ramanna visited our research labs and was impressed by the sophisticated equipment in our environmental radioactivity lab. Surveying the advanced spectrometers and measurement tools, he exclaimed, "Great to see your lab, Dr. Siddappa! I feel like I'm walking through the BARC labs! You have done a wonderful job". His heartfelt congratulations and a warm embrace filled me with profound pride and joy – it was the most inspiring moment of my life.

In March 1999, when I was appointed Vice Chancellor of Bangalore University by the Governor of Karnataka, one of my first visits was to Dr. Ramanna's residence (Fig. 6). He congratulated me warmly and encouraged me with, "Bangalore University is facing serious challenges. The Governor made a wise choice by appointing you to address them.

I am confident you'll resolve these issues and lead the university towards excellence. Wish you all the best". True to his word, he continued to support the university with both official and informal visits, inspiring me each time.

Reflecting on my 35-year association with Dr. Raja Ramanna, I am profoundly aware of how his enduring influence guided my journey from student to scientist. His mentorship and friendship have been invaluable, shaping not only my professional path but also my perspective on life.

The last time I met Dr. Raja Ramanna was just a couple of weeks before he took his final breath in Mumbai. We met at NIAS, shared lunch and discussed pressing issues in higher education. Reflecting on my tenure as Vice Chancellor, he commended the progress at Bangalore University, remarking, "I hope you will continue to reach higher positions to serve the country". His words still resonate deeply with me.

Fig. 6.   Dr. Raja Ramanna (left) and the author at the Bangalore University.

Swami Vivekananda once said, "Those who live for others live forever; others are more dead than alive". Dr. Raja Ramanna lived for others and his country, and because of this, he is still with us. His legacy lives on in the heart of every Indian.

Dr. K. Siddappa obtained M.Sc. degree in Physics with distinction from Mysore University and Ph.D. in Nuclear Physics from Andhra University and served in the Physics Faculty of Mysore University and Mangalore University for 33 years. As Founder Director of Microtron Centre, he established the Electron Accelerator Facility with sophisticated infrastructure. Using this facility, he carried out research in important inter-disciplinary areas covering Nuclear Physics, Radiation Physics, Material Science, Nano Science, Radiation Biology and Medical Science. The Centre has a nationwide users' community including BARC, ISRO, IISC, RRCAT (Indore), IIT (Madras), VECC Kolkata, IUAC Delhi and a host of universities and electronics industries. Department of Atomic Energy has recognised Microtron Centre as the centre of excellence and DST has recognised the centre as a national facility. UGC has recognised the centre as a centre with potential for further excellence.

Dr. Siddappa has guided a number of Ph.D. students and published more than 300 research papers in national and international peer-reviewed journals of repute. He has participated in a large number of International Conferences, chaired technical sessions and delivered invited talks. He has visited the USA, Germany, Sweden, Finland, Italy, Yugoslavia and other countries as a Visiting Professor and as a Visiting Scientist. Prof Siddappa has served as a member of Governing Councils and Governing Bodies of Nuclear Science Center, Delhi; Inter-University Center for Astronomy and Astrophysics, Pune; IIM Bangalore; ISEC Bangalore; Kidwai Memorial Hospital, Bangalore; and many other institutes of national character and in academic bodies of a number of universities. He has served in important high-level committees of UGC, DST DBT and DAE. As Vice Chancellor of Bangalore University, during 1999–2002, Prof. Siddappa introduced a number of novel and innovative programmes, streamlined the administration and brought discipline to the university. These programmes include the introduction of a women's study centre/department and a compulsory computer science course with practical training for all the postgraduate students for the first time in the country. A Bio-Diversity Park was developed on the 500 acres of the Jnana Bharathi campus of Bangalore University by planting more than 5 lakh saplings of Western Ghats species of rich diversity. The Bio-Park with rainwater harvesting and water bodies has a repository of Bio-Species with birds and animals. The Bangalore University under his leadership was accredited by NAAC with the highest, Five Star rating.

He is a recipient of several National and International awards, from UGC National Associateship to American prestigious society awards. Recently, he was awarded Sir M. Visvesvaraya Senior Scientist Award from Karnataka State Council for Science and Technology in 2011 for his lifetime significant contributions to science. Dr. Siddappa, as Honorary Director of JSS Foundation for Science and Society and now as President of Karnataka Association for Advancement of Science (KAAS), is engaged in strengthening science education at all levels, research in universities and in taking fruits of science to society.

Chapter 17

A Multifaceted Personality: Inspired Countless Scientists and Educationists

Karan Kumar

Entrepreneur, Business Coach & Education Enthusiast

Shruth & Smith Holdings, Bangalore, India

kk@karankumar.in

Although it is difficult to encapsulate his contributions, I feel privileged to join others in expressing my deep appreciation for this illustrious scholar and noble soul, who inspired thousands of researchers, academicians and policymakers. During my high school and college days, I often heard stories of great scientists, innovators and their contributions to the nation. Though I never saw them, their stories inspired young people like me to aspire to build a better nation. I was not only fortunate to see and interact with him but also blessed to work under his guidance during my formative years. He was none other than Dr. Raja Ramanna, the multifaceted personality, eminent scientist and father of India's nuclear program.

Indeed, it was an amazing coincidence for me, and I can never forget the day I met him for the first time at a seminar organised by the Professionals' Action Committee for Educational Reforms (PACER) Foundation, where Dr. Raja Ramanna served as a Patron along with Justice M. N. Venkatachalaiah in 1994. I was the coordinator for the PACER Foundation which had many academic leaders who were passionately involved in the policy matters of school and higher education. He was so humble and down-to-earth that he appreciated my views and respected our interaction during the seminar. To my great surprise, he invited me to meet him at his office – the office of the founder-director of National Institute of Advanced Studies (NIAS), an institution established by JRD Tata on the Indian Institute of Science (IISc) campus in Bangalore. Even to this day, I can recognise passionate and ambitious individuals, as I experienced this firsthand through Dr. Raja Ramanna's inspiring gesture in 1994.

With a reference letter from Prof. M. R. Holla (the then President of PACER Foundation and Principal of RV College of Engineering), I met Dr. Raja Ramanna in his chamber at around 10:30 AM and, without any hang-ups, we had a pretty long interaction exchanging our beliefs/views on educational substances. Later, at 12 noon, he suddenly rose from his chair and invited me to have lunch with him. We went to the dining hall though it was an early lunch for me. We sat across the table. Having seated opposite each other, he ensured that I felt completely at ease. While having lunch, he showed me how to eat drumsticks and sapota without using my fingers. When I came back to his chamber, to my greatest surprise, Dr. Raja Ramanna asked me whether I could work as a Research Associate in one of the Department of Science and Technology (DST)-sponsored projects that mainly dealt with education, research and management in higher education space. Though momentarily hesitant, I accepted his invitation. At around 1:00 PM, he called the principal investigator of the DST-sponsored project (Prof. C. V. Sundaram – formerly Director at Indira Gandhi Centre for Atomic Research (IGCAR), Kalpakkam) and introduced me. He not only directed Prof. Sundaram to engage me as his Research Associate but also requested him to train me for the first few weeks in the basics of research and methodology. Later, for the next 2 hours, I sat with Prof. C V. Sundaram, and he gave me many insights into what I should be doing and gave an assignment. I was asked to start the work the very next day, marking my first formal job after college. During the next 3 years, other than interacting with many senior scientists, I was associated with Prof. C. V. Sundaram as his Research Associate. Even to this day, though I have had many good coincidences in the last three decades, I can proudly say that meeting Dr. Raja Ramanna was an amazing coincidence and stands very special to me even to this day.

I should not fail to recognise his deep interest in music and the respect he had for musicians of all ages. He deeply enjoyed playing the piano. As witnessed by many, he strongly believed in the relationship between science and music, and it was his lifelong engagement. In my decade-long association with Dr. Raja Ramanna, not less than 50 times, I must have been invited with a few of his close associates to his house at RT Nagar in Bangalore to listen to him play the piano. He was an excellent host who always ensured we stayed for dinner.

In 1996, as the year was declared as "Year of Tolerance" by the United Nations (UN) and equally in recognition of several achievements of this world body, the United Nations 50th anniversary celebration was engrossed through a 2-day "International Summit" at the Choksi Hall in the Indian Institute of Science (IISc) campus, Bangalore. Mrs. Vijayalakshmi Thimmaiah (daughter of former Chief Minister of Karnataka Mr. Kengal Hanumanthaiah) was the chairperson of the Reception Committee. As part of the organising team, from the institution, Dr. Raja Ramanna called (being the co-chairperson of the organising committee) on the chairperson and informed her that he is deputing me to work in the secretariat of the organising committee on his behalf. Having worked in the secretariat of the organising committee for nearly 4 months, I had the great opportunity to meet and interact with many stalwarts including former Ambassadors, scientific administrators, secretaries of the Government of India and a host of other Diplomats. Dr. Ramanna used to personally participate at various stages and in all the organising committee meetings and through his connections, he was instrumental in having UN Secretary-General Mr. Boutros Boutros Ghali attend the Summit. Personally, it was a great learning and possibly the nicest gift received; I remain grateful to Dr. Raja Ramanna for this precious opportunity. In fact, all the guest rooms on the first floor of the NIAS were given to the guests invited to the Summit. Well, apart from the economic and social objectives for sustainable development that were set out by the UN, many topics including universal respect for human rights, global living standards and cross-cultural cooperation were deliberated in the Summit.

As Dr. Raja Ramanna was deeply interested and passionate about uplifting the school education, he wholeheartedly supported the initiative named PES-PACER Pilot Project that was chaired by Prof. M. R. Doreswamy (the Founder of People's Education Society (PES)). As he was very unassuming, Dr. Raja Ramanna never ever hesitated to attend even minor programs of the Government schools. Especially between 1998 and 2000, he must have attended so many programs of Government schools across the state of Karnataka. Being the coordinator of the PES-PACER Pilot Project which adopted 12 Government schools in and

Fig. 1. Inauguration of the new school building of Government School at Vajarahalli, on the outskirts of Bengaluru, 1998 (left to right): Dr. Raja Ramanna, Prof. M. R. Doreswamy, Chairman of PES-PACER Pilot Project, and Mr. R. Ashok, Local Member of Legislative Assembly.

around Bangalore for their academic and infrastructure development, I had the opportunity to accompany Dr. Raja Ramanna. As it was witnessed, he contributed more than 3 crore rupees from his MPLAD (Member of Parliament's Local Area Development) and other funds to upgrade the infrastructure of Government schools (Fig. 1).

He always introduced me as his young friend to his audience and a larger circle of associates. By taking full liberty, though we had a couple of generational gaps, I am happy to state that Dr. Raja Ramanna was not only an encouraging friend to me but also a tremendous source of comfort and motivation. As he has invited me on numerous occasions, Dr. Raja Ramanna was very fond of having lunch/dinner at the Bangalore Club (it must be the oldest club in the city of Bangalore). I noticed that he was very much influenced by Victorian-era society, tradition, culture and etiquette, and thus he loved to visit Bangalore Club again and again.

It was very unfortunate that we lost him on 24 September 2004. I still remember when he had once said that he had nothing to regret in life and that he was living his life to the fullest. Interestingly, in one of our interactions, he expressed his desire to die in Bombay (Mumbai), the city that had given him everything. In fact, as it was his wish, he was in Mumbai when he had his last breath.

In every facet of his life, he exemplified rational thinking, firm beliefs and uncompromising intellectual clarity. Above all, even in an informal conversation, he used to listen to people very prudently and quietly before coming up with his humorous anecdotes.

Though he was not around for me to mentor in the last 2 decades, I must simply place my sincere gratitude to Dr. Raja Ramanna as he has left his optimistic impersonations on me to be successful in my professional and social life as an entrepreneur, startup mentor, seed investor, business coach, education management and social activism.

Dr. Raja Ramanna was a Mentor, Leader, Nuclear Scientist, Philosopher, Accomplished Pianist, Towering Personality, Extremely Confident, Incredibly Humble, Prodigious Friend & a Complete Human Being – He was everything to me

---

Well, Dr. Raja Ramanna was one of India's most celebrated nuclear physicists. He was always looking onwards and upwards to contribute to national development with a sense of mission in all his capacities, which was all the more evident in his role as a minister of state for defence and a member of parliament.

As he was greatly interested in the "Philosophy of Science", "Sanskrit and Science" and "Science and Music", he will be remembered as a special scientist for whom both physics and the piano were music to his ears. There was an instance when Dr. Raja Ramanna passionately spoke of how his learning of music started at a very early age and in his high school days (i.e. Bishop Cotton's High School) he gave a piano concert. Likewise, apart from nuclear physics, I must say "music was his constant companion".

His centenary is the time to reorient ourselves to his exhibition of scientific temperament, progressive policies and a sense of belief in the younger generation for the progress of our nation. To me, being simple and humble, Dr. Raja Ramanna stands out as a shining example. Hats off to him!

Karan Kumar is formerly Research Associate at NIAS; Member of the Academic Council, Bangalore University (BU); Chairman of Bangalore University Task Force (BUTF); Member of the Executive Council, Visvesvaraya Technological University (VTU); Member of the Syndicate of Bengaluru City University (BCU); and presently, Member of the Board of Governors at BMS Engineering College and on the Governing Councils of several Academic Institutions.

Chapter 18

Dr. Raja Ramanna: A Mentor of Scientists and Nurturer of Scientific Progress of India

Dinesh Kumar Srivastava

National Institute of Advanced Studies, Bengaluru, India

dinesh.srivastava@nias.res.in

My profession is physics; music is my yoga.
– Dr. Raja Ramanna

A few months ago (2 September 2024), we had a very intense, invigorating day full of nostalgia, satisfaction at the fulfilment of dreams of detectors and accelerators, and visions of even loftier dreams, discussing nuclear physics' past, present, and future at the National Institute of Advanced Sciences, Bengaluru, as a part of a year-long celebration of the Birth Centenary of Dr. Raja Ramanna (1925–2004). We celebrated and rejoiced in the legacies of Dr. Ramanna, in the fervent hope that these discussions would have pleased the colossus among nuclear physicists in the country – who gave it a direction and presided over it for decades. We owe its present internationally competitive status to him, as he essentially created it and nurtured it to reach its present-day position of maturity.

I did not have the privilege of working immediately directly with Dr. Ramanna, as by the time I joined the Training School of the Bhabha Atomic Research Centre, one of his most prestigious and important creations, only in 1970, he was already very senior and Director of the

Physics Group of the Bhabha Atomic Research Centre. He did, however, visit us in the very beginning, gave a very motivating talk and left a lasting impression on us.

I vividly remember this visit of Dr. Ramanna. I was just 18 years old, and yet I asked him, "Sir, how will I know that the results I am getting from my research are correct". We were sitting on the roof of the then hostel for the Trainees, and the lights were rather dim. Looking back, I am sure that he must have suppressed a smile at my naivety. However, he replied in all seriousness to my stupid and infantile question that correct results will make a lot of things look consistent. He was not Sun to me, and I was deprived of his warmth which lucky persons such as Dr. V. S. Ramamurthy enjoyed. He was more like the Pole Star to me, who showed me direction throughout my career and did not let me deviate from my work.

I joined the Variable Energy Cyclotron Project in 1971. Dr. Ramanna was the Chairman of its Project Advisory Committee and guided its progress all along- removing all impediments.

My mentor, Dr. N. K. Ganguly, asked me to learn computer programming and to set up, by then standard, computer codes for the analysis of elastic and inelastic scattering of protons, deuterons and alpha particles from different nuclei using the optical model and distorted wave Born approximation for nuclear reactions. Around the same time, Dr. Bikash Sinha – a rather young but already quite famous nuclear physicist – visited us from King's College London, on the advice of Dr. Raja Ramanna. I started working on calculating the energy dependence of optical model potential – a subject of considerable interest at that time – for which Dr. Sinha had a plausible theory, which was not tested in detail till then. The computer program for those calculations took a very long time (they would take only a few tens of seconds or so on modern laptops!) on the CDC 3600 computer at Tata Institute of Fundamental Research, Mumbai.

Our group had been given a total of a few hours on that computer for every month. I quickly used up all the time allotted to us and had only started getting the hang of the calculations. All my colleagues – who could not use the computer any more – protested and suggested that I be debarred from using the computer. It was only after some time that I came to know that Dr. Ramanna – when informed of the problem – had a hearty laugh and allotted us sufficient time so that I could do my calculations and some more, without disturbing others. This also gave me the first publication of my life! And as luck would have it, I even got to present it at the

Department of Atomic Energy Symposium on Nuclear Physics at Panjab University, Chandigarh, in 1972, in front of him.

He often visited our labs in Bhabha Atomic Research Centre and in Calcutta on a regular basis to take stock of the progress and solve problems, if any – of any nature, administrative, scientific or technical. These visits of Dr. Ramanna used to fill all of us with enthusiasm, energy, and hope – in those desperate days – with very poor living conditions in Calcutta, extremely bad power supply and very poor industrial infrastructure – coupled with shortage of funds, especially foreign exchange, and technology denial imposed upon us post 1974. Our excitement, when he congratulated every member of the project from the Director to the workers such as cleaners, when we got the external beam in 1979 and utilization of the cyclotron started, is still fresh in our memory. He continued to visit us on a regular basis and listen carefully to all that we were doing and encourage us to aim higher (Fig. 1).

Fig. 1.   Dr. Raja Ramanna during one of his regular visits to VECC, Kolkata (~1985).

Note: Seated along the table: (Left to right) Dr. A. S. Divatia, Dr. Raja Ramanna, the author, Dr. Y. P. Viyogi and Dr. Kewal Krishan; C. K. Khasnabis is seen in the row behind.

Among many memories, the following specific memories stand out, when I think of Dr. Ramanna:

The first one is of the Summary Talks that he traditionally delivered at the end of the Annual DAE Nuclear Physics Symposium, one of his noble creations, continuing since 1957. In the early days, it used to be Nuclear Physics and Solid-State Physics Symposium – till the number of participants became too large and it had to be bifurcated. These symposia gave me friends across the country and remain the markers against which we recall the events of our scientific lives.

Dr. Ramanna used to make even very difficult topics look very easy and his Summary Talk used to be full of insights, humour, valuable observations and directions for future growth. He used to rejoice in the hard work and exciting results of colleagues and especially identify promising young researchers from across the country. He was quick to spot brilliant students and made a point of seeking them out and encouraging and enthusing them. And of course, he was an excellent communicator. He specifically chose these symposia as opportunities to create consensus on the creation and utilization of new facilities, and I will come back to this point.

Dr. P. K. Iyengar, his equally famous successor, famously called him an energiser. Nothing could be closer to the truth than this, as he remembered even the junior-most employees who were contributing to our programmes and rejoiced in the successes of his colleagues. He was also forever alert to openings and opportunities which could help his colleagues and nuclear research in the country in any manner. The recently concluded Symposium on Nuclear Physics at Indian Institute of Technology Roorkee (December 2024) had over 650 participants, about 40% of them were women and girl students.

While giving the Summary Talk at the above meeting, I had compared the growth of nuclear physics to the rise of cricket in the country. Till several decades after the independence, the cricket players belonged to royal families and rich clubs. The Indian Premier League changed all that – now we have cricket players from across the country and from all walks of life, and the vast potential of the youth of India is on display, not only in cricket but also in all sports and athletics, as well as in chess, and they are dazzling the world with their brilliance. These symposia took nuclear physics research out of select elite institutions. It has witnessed an exponential growth in the country, with participation from researchers from all corners and remote centres. It only goes on to prove Dr. Ramanna's conviction that if you create opportunities for the young people of India, they will prove to be second to none.

The second sacred memory involves the electrifying moment as he entered the jam-packed and over-flowing Central Complex Auditorium of the Bhabha Atomic Research Centre in Bombay – a few days after the Peaceful Nuclear Explosion experiment performed on 18 May 1974, when he returned from Pokhran to a tumultuous standing ovation lasting till our palms hurt. His response was still full of humour despite the enormous success achieved "silently", "mysteriously" and "miraculously". I still recall his amusing comment that the only worry he had was about how was he going to explain the loss of the cameras – installed in the pit in which the device was put and exploded – to the authorities as those were "capital items"!

The last and most sacred memory I have of him is that of his last visit to the Variable Energy Cyclotron Centre, Kolkata, to unveil a bust of Dr. Homi Jehangir Bhabha and to inaugurate the newly constructed Meghnad Saha Auditorium on our campus. The most delightful part of that event was an hour-long mesmerising, celestial and sublime piano recital by him, followed by a santoor recital by Pundit Shiv Kumar Sharma (Fig. 2).

Fig. 2. Left to right: Dr. Raja Ramanna, Dr. Bikash Sinha and Pundit Shiv Kumar Sharma at Dr. Ramanna's piano recital on the occasion of the inauguration of Dr. Megh Nad Saha Auditorium Kolkata (inset top left corner), 7 March 2004.

The two giants also played a short and delightful piece together, to the utmost delight of all of us. This was especially nice as we saw his facets as a great scientist, a great speaker and a great musician – within a span of about two hours. Little did we know that we were seeing him for the last time and that we would soon be instituting Raja Ramanna Memorial

Lecture – the first one of which was given by Dr. Swapan Chattopadhyay, associate laboratory director and distinguished professor at the Jefferson National Accelerator Laboratory, Newport, US, in June 2005. It was presided over by Dr. V. S. Ramamurthy. Since then, it has been delivered by world-renowned scientists, every year on the Foundation Day (June 16) of the Variable Energy Cyclotron Centre, Kolkata (Fig. 3).

Fig. 3. Dr. Swapan Chattopadhyay, Associate Laboratory Director and Distinguished Professor at the Jefferson National Accelerator Laboratory, Newport, US, delivering the First Raja Ramanna Memorial Lecture at VECC Kolkata in June 2005.

Years earlier, I had translated a long interview that Dr. Ramanna had given to Dr. Bikash Sinha, the then Director of the Variable Energy Cyclotron Centre for a Kolkata newspaper, *The Telegraph*, into Hindi for *Ravivar*, a prestigious Hindi magazine. Translating his words into Hindi, I realised the depth of his passion for science and music, and most importantly for the progress of our country and the alleviation of the poverty of its people. One specific remark that he made has stayed with me, "Nobel Prize is not the only criterion by which a scientist needs to be judged. They should also be judged by what they do to alleviate the suffering of masses using science and technology and to make their society secure and prosperous". He had also talked of nuclear energy and the extensive use of nuclear radiation in treating cancer, developing new crops, preservation of food and of nuclear techniques in improving the quality of life of masses.

The studies of basic nuclear physics in the country have progressed from simple beginnings using alpha, beta and gamma sources, neutron generators, a small cyclotron at Saha Institute of Nuclear Physics and a Van de Graaff generator at the Bhabha Atomic Research Centre, Mumbai (now upgraded to a Folded Tandem Ion Accelerator (FOTIA) to a K130 Variable Energy Cyclotron, a K500 Super Conducting Cyclotron and two Pelletrons with superconducting linac boosters. Our researchers now have access to Relativistic Heavy Ion Collider at Brookhaven, Large Hadron Collider at CERN Geneva and the Facility of Anti-proton and Ion Research, FAIR, at Darmstadt. India is an Associate Member of CERN, a part owner of the FAIR facility and a participant at the International Thermonuclear Experimental Reactor or ITER under construction in France. The condensed matter community has access to INDUS-1 and INDUS-2 synchrotron radiation sources, high flux neutrons from the Dhruva (built under Dr. Raja Ramanna's watch) and the recently refurbished Apsara reactors (the original Apsara reactor had been designed by Dr. Raja Ramanna) and will have a neutron spallation source in near future. In addition, our researchers have been participating in experiments at all major experimental nuclear physics facilities across the world individually and as a group. There is a phenomenal and exponential change in the energy of the projectiles and complexities and reach of detector systems during this period. The foundations for these international collaborations were laid by Dr. Ramanna who prepared the ground by enthusing us, building our capacity and preparing us to stand and participate in these endeavours as equals.

Let me add that one of the last acts of Dr. Homi Jehangir Bhabha, before he went on the fateful journey to Geneva, was to make a decision to build a Variable Energy Cyclotron in Kolkata. Years later, Dr. Ramanna was to persuade Mrs Indira Gandhi, to agree to buy a pelletron heavy ion facility for BARC/TIFR Bombay and build a multi-GeV synchrotron radiation source, for condensed matter studies as well as a Super Conducting Cyclotron.

The pelletron facilities, as mentioned earlier, at Mumbai and Delhi were upgraded with superconducting linear accelerators developed indigenously for heavy ions to study fusion–fission dynamics which among many other things, repeatedly confirmed the nucleon-exchange mechanism envisaged by Dr. Ramanna for explaining the mass distribution of fission fragments.

The accelerator physicists at Variable Energy Cyclotron Centre went on to build and commission a Super Conducting Cyclotron. They are also developing a radioactive ion beam facility. They designed several components for the Facility for Anti-proton and Ion Research (FAIR) at Darmstadt, which is partly owned by India! Several of them also moved to the Raja Ramanna Centre for Advanced Technology and additionally participated in the construction of the Large Hadron Collider and opened the doors for India to become a major participant in experiments at CERN and participate in the discovery of the Higgs boson and quark–gluon plasma, using detectors conceived, designed, developed and constructed in India. India is now an Associate Member of CERN.

The nuclear theory scene has progressed from simple Hauser–Feshbach and optical model calculations, and simple nuclear structure calculations to *ab initio* shell model calculations and nuclear reactions, and the use of density functional theory to study dynamics of fission of heavy and super heavy nuclei.

The contribution of our researchers to the theory of relativistic heavy ion collisions and quark–gluon plasma is both profound and extensive. Neutrino studies – especially neutrino–nucleus interactions – is yet another field where our researchers have made pioneering contributions. From the use of modest computer programmes, run using punch cards, we have graduated to using supercomputers and massively parallel processing computers, developed in India, for calculations and simulations and vast resources of grid computing spread across the world to analyse the data taken at CERN, for example.

This volume has touching and glowing tribute from many researchers and educators from various fields, who took guidance and inspiration from Dr. Ramanna to take up very difficult but transformative projects.

One of the things I learned from those contributions is that Dr. Ramanna extended helping hands to several institutes across the country and rescued them from perhaps imminent closure due to paucity of funds and raised them to a position of glory. During his short stay with the Defence Research and Development Organisation, he rescued it from the stifling clutches of the Union Public Service Commission, enthused its researchers by introducing merit promotion schemes, facilitated collaborations and introduced and initiated challenging programmes of missile, light combat aircraft, and nuclear submarine developments, which have all prospered and catapulted India to the position of a responsible world power.

One aspect of the contributions of Dr. Ramanna which is not very well known is the role he played in introducing radiation therapy, nuclear medicine and the production of radiopharmaceuticals for medical imaging in the country at nominal rates for our people. He also steered the programme research in high-yielding lentils, peanuts and other food grains using gamma radiations, as well as preservation of food grains and vegetables and sanitising of medical supplies.

My present organisation (NIAS) is truly a reflection of its founder, Dr. Raja Ramanna – who was a brilliant scientist, a great thinker, well versed in Indian and Western Philosophy, an adherent to human values and deeply interested in arts, music, paintings, architecture and the people of India. NIAS accordingly has multidisciplinary activities in education, sciences, climate, energy, security, philosophy, consciousness studies, conflict studies and poverty alleviation studies.

We lean back in satisfaction that the country and its important institutions are in competent hands, and we look towards a wonderful future. I would like to pay my tribute to one of the greatest sons of Mother India – who opened the doors to security forever for our country and to an ever-widening scope of research and development in all aspects of nuclear physics, nuclear energy and nuclear technology.

Prof. Dinesh Kumar Srivastava (born 1952) is presently Indian National Science Academy Senior Scientist and Honorary Visiting Professor at the National Institute of Advanced Studies, Bengaluru. He is now working on energy, environment, climate change, international collaborations and science outreach along with a continuation of his research on quark–gluon plasma.

He obtained his graduation from Allahabad University in 1970 and joined the Training School of the Bhabha Atomic Research Centre, Mumbai. He started working at the Variable Energy Project of Bhabha Atomic Research Centre in 1971 and retired as Director and Distinguished Scientist at the Variable Energy Cyclotron Centre, Kolkata, in 2016. Later, he continued there as Department of Atomic Energy (DAE) Raja Ramanna Fellow until 2019. He worked at National Institute of Advanced Studies, Bengaluru, as a Homi Bhabha Chair Professor, during 2019–2023. He is a Fellow of the National Academy of Sciences, India, and the Indian National Science Academy.

https://doi.org/10.1142/9789819814435_0019

Chapter 19

Ramanna: The Man and His Legacy[*]

B. A. Dasannacharya

Bhabha Atomic Research Centre, Mumbai

*Inter University Consortium for Department
of Atomic Energy Facilities, Indore*

adasannacharya@gmail.com

It is a privilege to be invited to speak during Dr. Raja Ramanna's birth centenary celebrations. I thank Dr. S. M. Yusuf, Director of Physics Group, Bhabha Atomic Research Centre, for the same. I am proud to add my few words of appreciation and express my deep respect for Dr. Raja Ramanna. His legacy is so widespread that it will be possible for me to give only a glimpse of the same in the time allotted to me. Almost certainly many of the things I describe would have been covered by earlier speakers. My apologies for the repetitions as well as deletions. However, I feel especially delighted to be addressing a young audience who are sometimes told that India hasn't been able to produce top-level scientists. I hope that by the end of the talk, that impression will have changed for the better.

---

[*] This article is based on lecture that was given by the author at Symposium on "Recent Progress in Nuclear Science & Technology" held at the Bhabha Atomic Research Centre Mumbai during 27–28 January 2025, to celebrate the birth centenary of Dr. Raja Ramanna.

Coming to my familiarity with Dr. Ramanna, I would like to go back to the first day I met him. The year was 1957 and the month I think was May. I saw a young Ramanna for the first time as he got down from a cycle rickshaw in front of our house at Banaras Hindu University. My father, then 61 and almost double his age, had retired from being the professor of physics there and had invited Dr. Ramanna to be the examiner of his Ph.D. student, K. S. Subuddhi, who had built an 800 kV Van de Graaff machine, arguably the first such device in India in 1957! It was a surprise for me to see such a young person being an examiner for Ph.D. But within a very short time, it seemed that I had known him for a long time. There was this special quality with Raja Ramanna that everyone felt at home with him even after a short meeting. Thirty years later, he autographed a photograph of himself for me when he retired from being the Chairman of Atomic Energy Commission and Secretary of the Government of India, Department of Atomic Energy (DAE), addressing me as a "dear and old friend" to my great delight (Fig. 1).

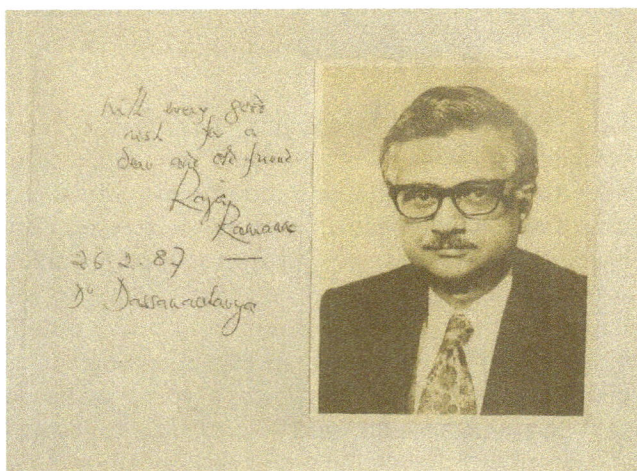

Fig. 1.   Autographed photograph of Dr. Ramanna for the author (1987).

I was 19, just completing my B.Sc. at the time of his first visit. This visit of RR was of great consequence to me in choosing my career path because my own plan to do M.Sc. in Physics and go abroad to do Ph.D. underwent a gearshift after his visit. That is a separate story for another day but if you prefer to count this as a minor legacy of Ramanna, I shall not object.

Dr. Ramanna's story has a parallel to this. But allow me to start right in the beginning. He started taking lessons in piano at a young age of six and developed keen interest in Western music. He gave a piano recital for the Maharaja of Mysore when he was twelve and received much appreciation and encouragement which he would be grateful for all his life. Then, in 1944, when he was 19 years of age, he was introduced to Homi Bhabha who was on vacation in Mysore. By then, Ramanna had been learning piano for more than a decade and had become very proficient at it. Music had become his first love; it continued to be a lifelong passion. However, this meeting had a profound influence on his career path. He had the following to say about this meeting: "My meeting with Bhabha would determine the course of the next several years of my life. But even as I looked forward to the future, I was aware that my youth and my childhood were in the past". Much later in life, he went on to write a book titled "Structure of Music in Raga and Western Systems", published by Bharatiya Vidya Bhavan in 1993.

Along with music, Ramanna had by then developed a keen interest in Nuclear Physics. After passing B.Sc. (Honours) from Madras University, he got admission to University College, London, to do Ph.D. in Experimental Nuclear Physics, fission physics in particular, and joined there through a Dorab Tata Trust fellowship. He obtained his Ph.D. degree in 1948 at the age of 23. Even before this, in 1947, Homi Bhabha was visiting London and the two met again. This time Bhabha offered him to join the recently established (1945) Tata Institute of Fundamental Research (TIFR), Bombay, and initiate low-energy nuclear physics research (experimental research in Cosmic Rays had already been initiated).

Ramanna accepted this offer and joined TIFR in 1949; he was just 24 years old. He was allotted one room for setting up his experiments and another for his piano. This shows how both Bhabha and Ramanna valued similar aspects of life. Bhabha was obviously impressed by his broad interests. Contrast this with the fact that Jagadish Bose had to be satisfied with doing his pioneering experiments on microwaves under a staircase at Presidency College, Calcutta, fifty years earlier. This would be unimaginable even today in the existing university system. The need for such a liberal approach followed by Bhabha completely embodied Ramanna's character and he, in his turn, passed it down to the Nuclear Physics

Division, Atomic Energy Establishment Trombay, now called Bhabha Atomic Research Centre (BARC), Mumbai, of which he was the Head. You may call this his legacy number two.

Dr. Raja Ramanna (RR) was a complete man with a holistically developed superior personality. Dr. M. G. K. Menon, a well-known Cosmic Ray physicist and a contemporary of Raja Ramanna described RR in the following terms: "Dr. Raja Ramanna was, in many ways, a towering personality with interests and accomplishments covering wide areas of nuclear science and technology, institution building, music, philosophy, human resource development, science policy, planning and administration". This is indeed high praise for the person being described, especially when one recognises that every quality has been carefully chosen by a person as exalted as Prof. M. G. K. Menon. I shall describe some of these.

Let us consider one part of nuclear science and technology and its planning and administration to start with. RR's personal area of research from the days of his Ph.D. was fission physics. With the installation of a 1 MeV Cockcroft–Walton machine at TIFR, he started experiments on neutron thermalisation along with fission experiments. These were continued at the Apsara reactor at Atomic Energy Establishment Trombay (AEET), now Bhabha Atomic Research Centre (BARC), after 1956. In addition, thanks to Homi Bhabha's contacts from his Cambridge days, he arranged for some of his early students such as P. K. Iyengar, G. Venkataraman, K. Usha, N. Umakantha and V. P. Duggal to be trained at Atomic Energy Canada Ltd, Atomic Energy Research Establishment (AERE), Harwell and France in using thermal neutrons from a reactor for doing experiments in Solid State Physics. Still later, when possibilities for doing condensed matter experiments with pulsed neutrons were showing promise, P. K. Iyengar, M. Srinivasan and others went to Italy and the USSR to learn about different types of pulsed neutron sources. This had implications for fast critical systems which have relevance in connection with building explosive devices also. Given Ramanna's comprehensive understanding of the field, his efforts at putting together personnel for all activities needed for Peaceful Nuclear Explosion (PNE) was a masterpiece. R. Chidambaram informs us in his autobiographical memoirs that Ramanna asked him to prepare the ground for a Peaceful Nuclear Experiment as early as 1967. This was after Homi Bhabha's death in an aeroplane accident in 1965 and seven years before the Peaceful Nuclear Experiment of 1974. It was a grand experiment involving generating top-class basic science results, front-line technologies and extraordinary planning and coordination. BARC today has world-class laboratories for high-pressure physics, thermal neutron scattering, computer simulation of atomic explosion devices and other relevant technologies. These facilities have been generated entirely by researchers largely trained locally. Hence, legacy number 3 is to select the best, train them to the full and trust them to deliver results.

The example given above consists of only a small part of the overall activities of BARC and DAE. Homi Bhabha had realised very early that if one wishes to launch a full-fledged programme to produce atomic energy and multifarious nuclear applications in industrial, medical, agricultural and many other areas, one could not depend on the existing educational system to provide properly trained personnel. Consequently, he gave Dr. Ramanna the responsibility of creating a training programme for scientists and engineers who would be brought up to the mark for starting their careers with various units of DAE. A scheme was put together under Ramanna's leadership to recruit about 150 graduates annually and employ them at the end of the training. Starting in 1957, this has produced more than 7000 S&T personnel including people occupying the highest position in all DAE units and many outside institutions. I am a product of the first batch of this school, and I think most of you also have graduated from this BARC Training School. I don't have to tell this audience that this has been arguably the most successful large science and technology programme for generating competent S&T manpower. This then is the fourth legacy of RR.

In relation to the totality of nuclear science and technology, I can do no better than to quote Dr. P. K. Iyengar, former Chairman of Atomic Energy Commission, India, and a long-term colleague of Dr. Ramanna, who said, "Out of the uncertain beginnings in the 1950s, if we have today achieved the status of a "developed country" in nuclear science and technology, it is in large measure a consequence of Dr. Ramanna's ideals, policies and efforts. He certainly leaves behind the proud legacy of a magnificent edifice of scientific achievements and attainments towards the country's energy and national security".

Some years after PNE, when Shri Morarji Desai was Prime Minister, Dr. Ramanna became the Scientific Advisor to the Defence Minister following Prof. M. G. K. Menon. Menon writes, "...He (Ramanna) however, accomplished a great deal during this (4 short year, 1978–1982) period-particularly in bringing into Defence Research and Development Organisation (DRDO) some of the culture introduced by Dr. Bhabha in the functioning of the atomic energy programme, this resulted in reduced bureaucracy; making the organization more scientist friendly, considerably improving recruitment and personnel policies, … , etc., … (and Menon mentions other issues). In many of these areas he was able to bring to fruition some of the initiatives that I had struggled with battling the bureaucracy. He also gave Dr. A. P. J. Abdul Kalam, now President of India, the opportunity to move from Indian Space Research Organisation (ISRO), after his success on the SLV-3 project, to head the Defence Research and Development Laboratory in Hyderabad responsible for the missile program". A. P. J. Abdul Kalam, about six years, 1978–1982 younger than Ramanna, said, "For us in the science and technology community, Dr. Ramanna was always a source of inspiration and guide". This is Ramanna's fifth major legacy.

The last legacy I shall mention is the institution which he brought up from its birth: National Institute of Advanced Studies, better known as NIAS, Bengaluru. To the best of my understanding, this place came up because of the feeling of Shri J. R. D. Tata that quite often very senior members of large companies were too confined in their knowledge base to their day-to-day responsibilities, leading to a narrow vision in general. J. R. D. Tata wished to see a place where a broader vision could be imparted to them leading to a healthier general growth of industry. Conversely, academicians could benefit by getting to know fields other than their own. Obviously, only a person with a similar vision backed with adequate experience was needed to take the concept forward. Not surprisingly, Dr. Ramanna was almost a unique person to take this position. And today, we have NIAS, the only institution of this kind in India.

In addition, he was responsible in a major way for piloting the establishment of the Variable Energy Cyclotron Centre at Kolkata in the seventies for nuclear physics research and the Centre for Advanced Technologies at Indore in the eighties for the development of accelerators, lasers, low-temperature physics and other techniques: legacies galore! The grateful nation named the Centre at Indore after him.

One can think of other legacies of Raja Ramanna but let us shift to some of his other personal interests. In his personal subject of interest, fission physics, he introduced the stochastic model of fission which theoretically describes fission of uranium through a process of random diffusion of nucleons between two incumbent fission product nuclei. This was followed further by applying this theory for heavy ion reactions, by him and his colleagues, including V. S. Ramamurthy, R. Subramanian, R. N. Iyer, S. K. Kataria and others.

He was interested in some deeper aspects of Mathematics, and it was from him that I was thrilled to know of Goedel's incompleteness theorem. For me, Heisenberg's Uncertainty Principle and Goedel's theorem were two pillars which for the first time, around 1930, introduced in physics and mathematics, respectively, the concept of "unknowability". It told me, philosophically speaking, the impossibility of knowing absolute reality: one only has a perception. This has been a guiding principle in my life regarding the nature of reality which, I think, is the central problem in anyone's life. Ramanna had put in much serious thinking on this subject.

I may further point out his deep interest in philosophy and Sanskrit. Swami Anand had told Gandhiji during the non-cooperation movement that if he wished to truly understand Bhagawad Gita, he should not read its translation by someone but instead translate it himself: the result was Gandhiji's Gujarati version which he called Anasakti Yoga. Ramanna took a similar approach to understanding a subject: when he wished to understand Mukundamala, an ancient Vaishnavite text, he translated it into English. My former colleague and lifelong friend K. R. Rao wrote in an article in *Current Science*, "He rendered into English, Mukundamala of Kulashekhara Alwar, a publication of the Gandhi Centre of Science and Human Values of Bharatiya Vidya Bhavan, Bangalore (1997). Although this work was published as recently as in 1997, interestingly, the Introduction to this translation by Ramanna dates back to 1974, the year, which saw him leading the Pokhran – 1 test. Ramanna wrote in the Introduction, 'It is unfortunate that many of the ancient poetry are so much associated with the ritualisation aspect of religion that their beauties are available only to a few who belong to the associated sect or caste and who in turn consider them as their sole property. This great Bhakti poem. ... is usually associated with the Vaishnavite sect; the sect itself was in the nature of Hindu reformist movement which rose to its zenith of influence under Sri Ramanuja.... It was a reaction against the agnosticism of the Bhuddhists and intellectualism of Advaita. ...'" He continues, "Social conditions have changed in the last thousand years and while we may have lost sight of the worldly problem referred to in the problem in the original sense, the greed for money and aimless seeking of ephemeral pleasures of life are so much around us, that the word "worldliness" (free translation of Samsara by Ramanna) can be interpreted as to refer to these ills". One would wonder if Ramanna had implied this broad vision whenever he referred to his own interest in yoga because a yogi is said to be aware of the ephemerality of all that is associated with this life, as stated in the other classic poem Bhaja Govindam or the greatest epic poem Bhagavad Gita. Concluding the Introduction, Ramanna notes, "Whatever be the thoughts contained in this inspiring poetry, it brings out the fact that no human activity reaches its full glory without "Bhakti", whichever way one interprets the word". In this one sentence, he seems to have encompassed many a concept such as Akaamyakarma (performing every action as a duty without seeking fruits of action) if not Sharanagati. One would do well to dwell upon the real meaning of the word "Bhakti" in this context!

This is my tribute, shraddhanjali, to Raja Ramanna, a major inspiration in my life, on his centenary celebration. An avadhani and an Indian von Humboldt, Raja Ramanna will always be remembered for his contribution to Indian life in a number of ways. Or maybe, I should end this with another quote from Dr. P. K. Iyengar: "Ramanna's more important legacy is his uncompromising belief in intellectual clarity and rational thinking in every facet of life, and his unwavering belief (which he inherited from Jawaharlal Nehru and Homi Bhabha) that the nation could progress only by embracing science and scientific thinking. The best way to honour his memory is not through eulogies, but by rededicating ourselves to his policies and belief".

Bibliography

1. Mahanti, S., Indian Scientists: Saga of Inspired Minds, in *Vigyan Prsar*, 2016, p. 156.
2. Menon, M. G. K., Raja Ramanna: A Brief Memoir, RRCAT Newsletter, reproduced from *National Academy of Science Letters*, 28, Nos. 7&8 (2005).
3. K. R. Rao, Raja Ramanna – A personal tribute, *Current Science*, 87, 1152 (2004).

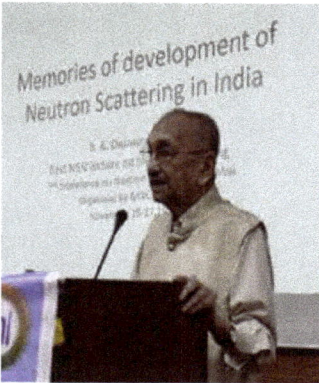

Dr. B. Anantha Dasannacharya joined the Atomic Energy Establishment, Trombay (now Bhabha Atomic Research Centre, (BARC)), after doing B.Sc. from Banaras Hindu University and received M.Sc. and Ph.D. degrees by research from Bombay University under the guidance of Dr. Raja Ramanna and tutelage of Dr. P. K. Iyengar and Dr. G. Venkataraman while working at BARC. He worked under Prof. B. N. Brockhouse at Chalk River Nuclear Laboratories, Canada (1961–1967), on neutron scattering by germanium and liquid argon. He was an expert scientist at Philippines Atomic Research Centre (1966–1967). He was a Guest Scientist at Institute fur Festkorperforschung, Julich, Germany (1973–1975); Royal Society-Indian National Science Academy exchange Fellow at (then) Rutherford Appleton Laboratory and Atomic Energy Research Establishment, UK (1981); and Indian National Science Academy–Japan Society for the Promotion of Science Exchange Fellow (1994). He worked in various positions at BARC such as Scientific Officer (1958–1986); Head of Nuclear Physics Division (1987–1990); Director of Solid State and Spectroscopy Group (1990–1996) and Additional Director, Multi-disciplinary Technology Group (1993–1996). He became Director (1995) of Inter University Consortium for Department of Atomic Energy Facilities (IUC-DAEF) (now called UGC-DAE Centre for Scientific Research), Indore.

https://doi.org/10.1142/9789819814435_0020

Chapter 20

Dr. Ramanna: A Leader with a Difference

Anitha Kurup

National Institute of Advanced Studies, Bengaluru, India

bkanitha@nias.res.in

As a not-so-young girl in my early thirties, I was looking for a job opening in research soon after submitting my Ph.D. thesis, which took a good five years. During those years, girls like me had very little space to negotiate to get married late. Hence, the challenge to navigate my Ph.D. with marriage, a young child and more importantly the anxiety at the end of submission of not being employed and a full thirty years of age.

It was the mid-nineties when Bangalore was relatively more beautiful, and the traffic was not so much of a challenge as it is now. Having taken the first job in an NGO-run training institute not very far away from home, as a research associate in January 1994, I quit in November the same year due to differences with the organisation's leadership over the credits for an innovative training module series my friend and I had developed to train the first women entrants to the gram panchayats. This work was in association with the Department of Women and Child Development, Government of Karnataka. Not surprisingly, we were young and gave little attention to what it would mean to quit our first job in 11 months and the repercussions it would have on our careers. No consultations but perhaps the arrogance of youth and the sheer confidence that we will find ourselves a job soon.

My friend and I had met at this training institute, but the friendship and trust we shared within a year was rather unusual. It was then that both of us looked up advertisements and started applying for work. The understanding between us was that both of us would apply for all the jobs that we come across which we like, and whoever gets it will help the other find her job.

It was the first week of December 1994, when both of us for the first time entered the National Institute of Advanced Studies campus through the Indian Institute of Science gate. It gave me a great deal of pride that I was driving through this esteemed campus that I aspired to enter as a little child. Driving through the small campus of NIAS at the far end of the IISc campus (a good two kilometres away), I was struck with the splendid garden and the variety of flowers I had not seen in any other campus before. We walked to the reception, and we saw a small, cute dog surveying the surroundings and giving us a look like what are you guys doing here. For a second, I wondered how this dog strayed into the campus that looked so well kept. And then I heard this firm friendly voice, "Hey Chin-Chin, come here".

I looked up to see who this distinct voice belonged to, and to my surprise stands this man whom I had admired, not even hoped to see nor meet in my lifetime – Dr. Raja Ramanna. He said, "Come in, whom are you looking for", and I said, "I have come to drop my CV for an advertisement I have seen in women studies", and he asked me to check with the administrator and that he would help me.

We dropped our CVs and on our way home, I told my friend there is no way I can travel to this place for work every day; it just is so far away. However, I attended the interview, which is a story for another day, but guess who were the experts on the interview panel? Two other stalwarts I did not think I would ever meet in my lifetime – Prof. M. N. Srinivas and Prof. R. L. Kapur. Both the names I had heard about and read their work but did not have an inkling I would get to meet them and subsequently, work with them so closely.

The setting of NIAS – a multidisciplinary institute that trained leaders from the public and private sectors – stands distinct from any traditional higher education institution. An institute that brought together experts from extremely different fields together on one hand and aspiring young professionals in the same fields on the other and created a space for co-creating and learning was like beyond the ordinary and indeed very special. The air of informality and the sense of freedom and autonomy was perhaps what attracted me and many other younger scholars to this institution, where the salaries during those days did not in any way match those provided by universities in the country. In fact, it was only a fraction of what perhaps individuals with similar professional credentials were paid. Forgetting about the distance, I decided to join NIAS when I was selected for the job.

What excited me was that we had the privilege of having lunch every single day with Dr. Ramanna, Prof. M. N. Srinivas and Prof. R. L. Kapur at the same table. These lunch table discussions were across a wide range of topics, and we got homemade food prepared by Mr. Krishnan and Mr. Satyamurthy. I remember I went home on the first day and told my family, "Guess with whom I had lunch today? Dr. Ramanna and Prof. M. N. Srinivas and Prof. Kapur. And ... I will continue taking lunch with them every single day from now on". The sense of pride and awe that engulfed me as a young professional starting a serious career in an academic space where the research agenda was defined by the scholars, not restricted to one's super specialisation had no bounds. It was like a dream institution, created for outliers like us, who disliked the traditional university setup where one continued to teach the same courses for the duration of one's service that could span over 3 decades. I was happy that I was free from the shackles of long teaching hours to students on a campus where the washrooms were a nightmare for me as a student. The buildings were old and unclean, and it would have been extremely difficult to keep my motivation day in and day out.

The National Institute of Advanced Studies (NIAS), founded by JRD Tata and Dr. Ramanna, set an exemplary academic environment with an attractive campus which is a delight for anyone visiting NIAS. The academic environment was distinct, where discussions and dialogues flowed easily between experts and scholars in a non-hierarchical style that laid the ground for the generation of new ideas and built the confidence in young scholars as we etched out careers. What made NIAS different from the other institutions was that this non-hierarchical style extended to the administrative staff and our blue boys. Dr. Ramanna led by example through his interactions with all irrespective of what levels each one of us occupied in the institute. For many of us who came from traditional educational institutions and universities, this was a marked difference in the academic environment which was like an ideal situation. It is perhaps because of this that many of us who started our careers here are an absolute misfit in any other higher education institution.

Wednesday Discussion Forum Followed by the Vada Session

Walking in for the first Wednesday discussion, was this man with long curls, speaker of the day, Sundar Sarukkai who pulled out a chair from the audience and said what is this? And someone replied it is a chair. "Who says this is a chair?" How did it get this nomenclature? Can I call it something else? And there went on a series of questions posed by him whom I later got to know is a philosopher. Hearing this for the first time, I wondered how I could make sense of something like philosophy that questions what is obvious. And there I got my introduction to the subject. I slowly realised that NIAS had scholars from different fields, and they engaged in interdisciplinary work for the first time.

The Wednesday discussion forum was a unique platform at NIAS created under the leadership of Dr. Ramanna. The forum provided a space for scholars and experts to present ideas/work in progress and derive feedback from scholars across different disciplines. The tea session with vada as an incentive provided one to carry over the discussion. The delicate balance of informality and confidence through which this session was conducted and the mutual respect shown to scholars across disciplines allowed one to engage in a free-flowing dialogue which was enriching and provided a fertile ground for new ideas thereby opening opportunities to young scholars and experts to gain knowledge of other domains. It is this forum that I can say with conviction gave rise to several important strands of research pursued at NIAS which made NIAS distinctly different from standard centres of higher education.

This tradition was carried out over the decades and has perhaps been a space that gave rise to new ideas. In one of these discussions, Dr. Ramanna said, "Oh you are studying women in Karnataka, do you want to study women in the Indian Institute of Science next door". Taking a cue, I embarked on one of my areas of research namely women in STEM and the presentation at the Wednesday discussion forum subsequently got Dr. Ramanna wondering, if the data I was presenting was correct. For the first time, I stated that they were less than 3% of the faculty of IISc. And he said- no way Anitha you have got it wrong, I see so many more young girls in the campus now than a decade ago. To which I had to reply, "Yes, the number of girls in the campus must have increased but it is not reflected among the faculty", thereby widening the scope of understanding the intersection of gender and science. Today, the work of NIAS on

Women in STEM has had an influence on policies that govern increasing the diversity in the STEM disciplines. I was the expert member of India on Diversity, Equity, Accessibility and Inclusivity and was invited speaker for the agenda item on *"Diversity, Equity, Inclusion, and Accessibility in Science & Technology (S&T)"* during the first *G20-Chief Science Adviser's Roundtable (G20-CSAR)* that was held on *28th–30th March 2023 at Gandhinagar, Gujarat, India.*

Along the vision of NIAS and Dr. Ramanna, who were keen that NIAS should train leaders for India and the world, the Education for the Gifted and Talented (EGT) was initiated in the year 2011 under the guidance of Prof. V. S. Ramamurthy and led by me where NIAS embarked on developing a National Programme to identify and mentor the gifted children in the age group of 0–18 years. The programme aimed to identify young leaders through this programme and provide mentorship till the age of 18 years. The programme developed and standardised India-based protocols to identify and mentor the gifted among urban, rural and Adivasi communities in Science, Technology, Engineering and Mathematics (STEM), social sciences, humanities and arts. To increase the outreach, a hybrid model to establish Advanced Learning Centres (ALC) in schools has been developed that provides mentoring through weekend classes, summer/winter workshops and one-to-one mentoring from 6th standard to 12th standard. A pilot project with 15 Army Public Schools across India demonstrated the positive impact of this hybrid programme. NIAS is currently conducting the hybrid model of ALCs in 30 more Army Public Schools across the country on a subscription model.

In addition, the current research project for the years 2023–2026 NIAS EGT is developing pilot projects of the hybrid ALCs in the Navodaya Schools, Kendriya Vidyalaya and schools for the urban poor. We propose to develop learning rubric to track the learning of the children in the ALCs. A total of 4000 teachers – urban and rural – from different states were trained. A Teacher Training Manual (English and Kannada), mentoring frameworks and resources to support the gifted children, has been developed. This year, NIAS has identified 1500 gifted children across schools in India and is conducting sessions and supporting 50 schools to establish Advanced Learning Centres in these schools.[1]

---

[1]For more on the programme, visit www.prodigy.net.in. See also: http://nias.res.in/professor-and-dean/anitha-kurup.

Fig. 1.   ALCHEMY 2019 (Author with the participants).

Alchemy 2019, a display of student projects at NIAS was held in collaboration with IISc (Fig. 1). On the same day, IISc open day was held that year. There were more than 2000 footprints that visited NIAS to see the display of Alchemy – An exhibition of the student projects who were part of the NIAS Advanced Learning Centres – A mentoring programme of NIAs education for the gifted and talented.

Associates Evening

Another hallmark feature of NIAS was the Monthly Associates Evening organised typically during the last Friday of the month. The introduction of this by Dr. Ramanna, as an important programme of NIAS, was to bring together important leaders, thinkers and professionals of the city together at NIAS and discuss topics that are important to the city, state, nation and the world. Speakers from across the country and sometimes even international speakers delivered a lecture followed by discussions and more importantly dinner with wine. I remember distinctly when Dr. Ramanna had told Mr Grover (a well-known owner of a vineyard) in a casual conversation to sponsor the wine for the Associates' evening, which he readily agreed. Jokingly, Dr. Ramanna mentioned, "You know ladies the wine was sponsored keeping you in mind, so you better stay for the Associates Evening". We were a very small group and when one of the women mentioned that it will be late to get back home, he turned and told

Maj. General M. K. Paul, who was the administrative head, "Please arrange transport for the women, we do not want them to miss out on the Associates Evening". Though NIAS was even then struggling for funds, Dr. Ramanna made it possible to provide transport to build a sense of belonging among all of us who were part of the NIAS family. Many of us were still beginning our careers. We were thrilled to listen to eminent speakers that Dr. Ramanna attracted by not only his stature but also his charm and his wide network.

Redefining Academic Environment

Dr. Ramanna in his usual style redefined the academic environment at NIAS which gave the scholars the liberty to define their own research interests and tied that to the responsibility of raising the resources to pursue the same. Bringing together the unique experience of meeting the country's leaders every year through the flagship programme for leaders and executives from the public and private sectors, the regular Wednesday discussion forum where research in the works was discussed and the comments and feedback greatly improved the ideas, and the Associates Evening every month that brought together more local audiences from Bengaluru city to again interact and carry out dialogues sparked by the lecture/art performances made NIAS an enviable place of work. We chose to work here despite the challenges of salaries back then, the uncertainty of sustained funding and the joy of embarking on research ideas not often within the confines of one's own specialisation. What more can one ask for? Looking back, we value the experience we had within this campus and undoubtedly pay tribute to Dr. Ramanna who set the stage for an institute such as NIAS.

Dr. Anitha Kurup is a Professor and Dean School of Social Sciences and Head of the Education Programme at the National Institute of Advanced Studies (NIAS), Bengaluru, India. She leads the National Gifted and Talented Education Program in India anchored at NIAS. The programme was initiated by the office of the Principal Scientific Adviser, Government of India, in 2011. She and her team have developed Indian-based identification protocols and mentoring mechanisms for the gifted children in India in the age group 3–18 years for different populations – rural, urban and Adivasis. She has developed successful multistage multi-level mentoring models for the gifted children in India. Prof. Kurup was part of the national team that successfully advocated for the education of the gifted and talented resulting in its inclusion in the National Education Policy 2020 of India. Her research career spanning over two decades is marked by her passion and motivation to undertake research in critical areas, hitherto unexplored within the Indian subcontinent. The hallmark of her research career has been the innovation methodologies adopted for large scale research studies questioning existing theoretical frameworks to find solutions to the real-world problems. Thus, her contribution to research stands the test of time and has made a critical contribution to the growth of the educational field in India. Her significant contributions were to gender and science, higher education and education for gifted and talented. She has several publications to her credit.

Chapter 21

Dr. Raja Ramanna and Ancient Indian Traditions

Sisir Roy

National Institute of Advanced Studies, Bengaluru, India

sisirnias@nias.res.in,

sisir.sisirroy@gmail.com

Dr. Raja Ramanna, an internationally renowned nuclear physicist and founder and Director of the National Institute of Advanced Studies (NIAS), had deep interests in Western classical music as well as in ancient Indian traditions. He published the book *Years of Pilgrimage: An Autobiography* in 1991 where he discussed the philosophy of music and its deep impact on the human mind. The effect of music on the human mind has become a fascinating subject in the field of neuroscience and has drawn large attention to the scientific community at the beginning of the 21st century. Here, he felt that the art of making music is not merely entertaining but a branch of Yoga which can take the mind to a higher level of consciousness. Apart from the publication of this book, Dr. Ramanna published a few thought-provoking papers on ancient Indian logic as well as Vedantic wisdom and epistemological issues in quantum theory.

It is worth mentioning that in his paper *Logic: Ancient and Modern/ Scientific*, he dealt with very abstract concepts of Indian logic for example *Saptabhangivada* in Jain Philosophy and "Catuskoti" in Buddhist paradigm. Essentially, the Jain logic is seven-valued and Buddhist one is four-valued logic in contrast to Boolean or two-valued logic used in modern science and technology. As far as the understanding of the present author goes, Dr. Ramanna was very much inspired by the developments of quantum theory and the measurement problem and their epistemological issues in the 20th century to study ancient Indian logic. The behaviour of microscopic entities such as electrons, protons and photons is governed by a logic which is also non-Boolean.

Dr. Ramanna published another fascinating paper *Moksha: A Critique* relating to new discoveries of science and the revised definition of Moksha. He claimed that the old saying of the Vedanta that all minds are entangled might be analogous to quantum entanglement. Of course, it needs to be critically analyzed in the light of modern quantum theory.

Dr. Ramanna started NIAS with great personalities such as Dr. R. L. Kapur and Dr. B. V. Sreekantan. Dr. Kapur was a very well-known psychiatrist. He was interested in issues connecting spirituality and psychology for many years. He started his investigations of whether yoga can influence internal mood states and their reaction to stress. The reason was whether and how the emotional reactions could be investigated objectively. Dr. Kapur's book *Another Way to Live* opens new vistas in this direction. Dr. B. V. Sreekantan, former Director of Tata Institute of Fundamental Research, Bombay, and an internationally acclaimed experimental physicist, joined NIAS to build up a truly interdisciplinary institute to fulfil the dream of Dr. Raja Ramanna.

Dr. B. V. Sreekantan and Interdisciplinary Approaches in Science

Since the very inception of NIAS, Dr. Sreekantan started to investigate and inspire young faculties about the role of interdisciplinarity in science. His main emphasis was to study the subjective elements of consciousness and cognitive activities objectively. This is in the same spirit as Dr. Kapur and in line with that of Dr. Ramanna's approach. He became involved in the debate on matter, mind and consciousness. As a physicist, he was intrigued by the weirdness of quantum theory and the role of psyche. He played a very prominent role in starting a department in consciousness studies programme which is unique in this country. With his encouragement, research on animal cognition and consciousness became one of the active areas of research in NIAS.

Dr. Sreekantan along with the young faculties arranged many international conferences and invited great personalities to NIAS to understand the nature and role of consciousness in living and non-living beings. Later in his life, he became more interested in dialogue between scientists and traditional scholars rather than the usual workshops or conferences. As a result of this, the book *Understanding Space, Time and Causality: Modern Physics and Ancient Indian Traditions* written by Dr. Sreekantan and Dr. Sisir Roy (the present author) emerged. It covers some of the pertinent issues both in science and the ancient Indian traditions which were a part of the dialogue between scientists and traditional scholars.

Modern Science and Ancient Indian Wisdom
On Logic: Ancient and Modern/Scientific:

Dr. Ramanna published one paper on ancient Indian logic, especially on Jain and Buddhist logic. In Jain philosophy, the *Saptabhangivada*, the seven predicate theory may be summarized as follows:

The seven predicate theory consists of the use of seven claims about sentences, each preceded by "arguably" or "conditionally" (*syat*), concerning a single object and its particular properties, composed of assertions and denials, either simultaneously or successively, and without contradiction. These seven claims are the following:

1. Arguably, it (that is, some object) exists (*syad asty eva*).
2. Arguably, it does not exist (*syan nasty eva*).
3. Arguably, it exists; arguably, it doesn't exist (*syad asty eva syan nasty eva*).
4. Arguably, it is non-assertible (*syad avaktavyam eva*).
5. Arguably, it exists; arguably, it is non-assertible (*syad asty eva syad avaktavyam eva*).
6. Arguably, it doesn't exist; arguably, it is non-assertible (*syan nasty eva syad avaktavyam eva*).
7. Arguably, it exists; arguably, it doesn't exist; arguably it is non-assertible (*syad asty eva syan nasty eva syad avaktavyam eva*).

There are three basic truth values, namely, true (t), false (f) and unassertible (u). These are combined to produce four more truth values, namely, three-valued logic. Though, superficially, it appears that there are only three distinct truth values, a deeper analysis of the Jaina system reveals that the seven truth values are indeed distinct. This is a consequence of the "conditionalising" operator "arguably" denoted in Sanskrit by the word *syat*. This Sanskrit word has the literal meaning of "perhaps it is", and it is used to mean "from a certain standpoint" or "within a particular philosophical perspective".

Formal logic, so far as we know, originated in two and only two cultural regions: in the West and in India. One country in which logic particularly flourished between the two World Wars was Poland. A new chapter in the study of Indian philosophy began with the appearance in Warsaw of Stanislaw and his establishment of the Oriental Institute at Warsaw University in 1932. In a short time, the Indology Department of the Oriental Institute became an important centre of research on the history of Indian, and especially Buddhist, philosophy. One of the topics of research was the study of Indian logic. They studied Jain logic extensively and the approach was not to compare Indian and modern logic to examine individual differences along with similarities; they evaluated Indian logic from the standpoint of modern logic to determine what is at all logical in the way we understand it now from the Western perspective. Soon, they realized the seven-valued nature of Jain logic in terms of the current understating of Western logic.

The Buddha, like his contemporaries, also made use of the "four corners" (*catuṣkoṭi*) logical structure as a tool in argumentation. According to Jayatilleke, these "four forms of predication" can be rendered as follows:

1. S is P, e.g. *atthi paro loko* (there is a next world).
2. S is not P, e.g. *natthi paro loko* (there is no next world).
3. S is and is not P, e.g. *atthi ca natthi ca paro loko* (there is and is no next world).
4. S neither is nor is not P, e.g. *n'ev'atthi na natthi paro loko* (there neither is nor is there no next world).

The Buddha in the Nikayas seems to regard these as "the four possible positions or logical alternatives that a proposition can take". Jayatilleke notes that the last two are non-Aristotelian.

Aristotelian logic is a deductive reasoning method that uses syllogisms to reach conclusions, while Boolean logic (which we use in modern science, say, in the case of computers) is a type of algebra that uses logical operators to calculate results as either true or false. The key difference between traditional (Aristotelian) and modern (Boolean) categorical logic is that traditional logic assumes that category terms all refer to actual objects. Modern logic does not make the existential assumption.

In the case of quantum theory, the logic is non-Boolean and three-valued. This is related to the superposition principle in quantum theory. For example, in the case of the famous Schrodinger's cat paradox, the cat is dead, living or in a superposed state of living and dead. According to a Boolean framework, there will be either true or false; hence, in the quantum domain, non-Boolean nature is prevalent. I presume Dr. Ramanna became interested in understanding the logic in the quantum domain and that may be one of the reasons he took much interest in ancient Indian logic such as Jain or Buddhism which are essentially non-Boolean. It is to be noted that a much-generalized framework for logic has been discovered by the end of the 20th century called "paraconsistent logic".

Paraconsistent logic is a way to reason about inconsistent information without lapsing into absurdity. In a non-paraconsistent logic, inconsistency *explodes* in the sense that if a contradiction is obtained, then everything else obtains, too. Someone reasoning with a paraconsistent logic can begin with inconsistent premises – say, a moral dilemma, a Kantian antinomy or a semantic paradox – and still reach sensible conclusions, without completely exploding into incoherence. It is shown that the above-mentioned quantum logic or Buddhist one of seven-valued Jain logic (??) can be consistently described within the framework of paraconsistent logical framework.

In *Moksha: A Critique*, Dr. Ramanna discussed some key concepts in Vedanta and quantum theory such as quantum entanglement and interconnectedness. To understand this issue, we need to analyse critically the concept of non-locality and quantum entanglement. In the case of non-locality, as described in the famous EPR (Einstein–Podolsky–Rosen) paper in 1935, two things should be noted carefully, i.e. preparation of the state, say, the state of two electrons or two photons, and then study their behaviours separated at large distances. There you will see that two entities are connected but not in a causal manner. This connectedness has been observed experimentally over a large distance, say, one on the Earth and the other one on a satellite. Usually, people say that all minds are connected according to Vadanta and hence it is similar to quantum entanglement. This seems to be a very simplistic statement. However, the concept of connectedness is not valid in Advaita Vedanta since you need duality for any kind of connectedness. It may be possible in other variants of Vedanta. Now, the most crucial question arises regarding the preparation of the state so that things are "being as connected". One possibility is that if two entities in the universe are prepared to be in a state of mind (through Yogic practices) with the same stage of consciousness, one may think of connectedness as pure consciousness is considered to be a substratum of all the entities. In Buddhist philosophy, the doctrine of "Pratityasamutpada" or "dependent co-arising" one can think of a Network popularly known as "Indra's Net" where all the entities are supposed to be connected through a network through Indra's magical power. But we need to understand the conditions under which this kind of interconnectedness manifests which is similar to the procedure of "preparation of the state" in the microscopic domain. Let me finish the discussion with a story widely circulated: "Asanga, the stepbrother of a famous Buddhist scholar, at his very early age went to a deep forest to meditate for Nirvana. Even after twelve years of meditation he did not achieve anything and became disappointed and came out of the forest. Then he saw a dog fully covered with maggots all over its body. He thought the dog would die if he was not able to take away these maggots. However, he thought that then the maggots would die. So, he took one big leaf from the nearby tree put away the maggots one by one and cut his (own) flesh so that the maggots would not die. Then he saw suddenly the dog disappeared and he got enlightened". The message is that unless all the entities in the universe are being enlightened, one will not get Nirvana or enlightenment. Dr. Ramanna talked about this kind of collective enlightenment or Moksha of humanity but one has to reach such a state so that the connectedness will manifest.

Prof. Sisir Roy is a theoretical physicist, and his main field of interests includes foundations of quantum theory, theoretical astrophysics and cosmology, brain function modelling and higher-order cognitive activities as well as ancient Indian traditions. He is currently a Visiting Professor at the National Institute of Advanced Studies and has been awarded the prestigious Senior Homi Bhabha Fellow by Homi Bhabha Fellowships Council, Mumbai, for a period of two years from June 1, 2018. Since November 2014, he worked as a Raman Pai Chair Visiting Professor at NIAS till May 2018.

Prof. Roy served as a Professor of Physics and Applied Mathematics Unit, Indian Statistical Institute, Kolkata, from 1993 to 2014. He joined as faculty in the same institute in 1980. He did his Postdoctoral research with Prof. Jean Pierre Vigier, Institute Henri Poincare, Paris, from 1986 to 1987 on foundations of quantum theory. He worked with Prof. Rodolfo Llinas (New York) on Neuroscience with Prof. Menas Kafatos, George Mason University, USA, and Prof. Ralph Abraham, University of California, USA. He visited many US and European Universities as a Distinguished Professor. He published more than 250 papers in international journals and 18 research monographs by Springer, Kluwer Academic Publishers, World Scientific Publishers, etc. His recent books include *Decision Making and Modeling Cognitive Function* (Springer, 2016), *Demystifying the Akasha: Quantum vacuum and consciousness* (Epigraph, New York, 2011) jointly with Prof. Ralph Abraham (USA) and *Understanding Space Time and Causality: Modern Physics and Ancient Indian Traditions* (Rutledge Taylor and Francis, NY and London, 2019) jointly with Prof. B. V. Sreekantan.

Chapter 22

An Old Monk: Reminiscences of Raja Ramanna in NIAS, 1996–2004

Anindya Sinha

National Institute of Advanced Studies, Bangalore, India

asinha@nias.res.in

anindya.rana.sinha@gmail.com

> *I am enough of the artist to draw freely upon my imagination.*
> *Imagination is more important than knowledge. Knowledge*
> *is limited. Imagination encircles the world.*
> Albert Einstein, 1929

I think Einstein's rather profound description of himself fits RR, one of my heroes in my professional life, perfectly as well. Years may have gone by but RR – as Dr. Raja Ramanna was known to us all at NIAS, the National Institute of Advanced Studies – continues to be my hero. One usually chooses an idol by recognising – and it could not be truer than this here – that one could never be him although, deep down, wistfully, one did wish that he could. It is that simple but also that impossible.

My earliest and most abiding memory of RR is of him, sitting in the audience, in a semi-circular arrangement, in what was then the lecture hall of the institute, currently the main dining hall for institutional use during conferences and meetings. It was a cool October day in 1996, and he was listening to me delivering an invited lecture. I was arguing gently, but firmly, about why I thought my beloved bonnet macaques – those somewhat pale-faced monkeys with long tails, ubiquitous in the Bangalore of those days, rarely seen today – were indeed "conscious" beings. Sitting in a cane chair – only some of which continue to hold their place in the NIAS of today – clothed in his characteristic white shirt, black trousers and a grey coat, dignity personified, RR seemed relaxed, his glasses glinting in the reflected light, a bemused smile on his otherwise serious face, as he listened attentively and thoughtfully to me, his hands occasionally running through his salt-and-pepper hair. But not a hair was out of place. While listening, he would sometimes extend his hand, almost absentmindedly, and gently stroke his close friend and canine companion, Chin-Chin, a personable Lhasa Apso, sitting faithfully by his chair, occasionally looking at me intently. It almost seemed Chin-Chin knew exactly where his friend's attention lay.

At the end of the talk, there ensued an intense debate. But while I found Prof. B. V. Sreekantan, the eminent physicist and former Director of the Tata Institute of Fundamental Research in Mumbai, who had invited me to deliver the talk, deeply engaging with the academic aspects of what consciousness was and how it could be defined, RR seemed to be fascinated by the monkeys themselves and what my studies on their consciousness had revealed to me about their lives. His questions were simple, straightforward, searing in their intensity but delivered quietly. I just could not believe that I was addressing one of the greatest legends of 20th century India, called a "towering personality" by Prof. M. G. K. Menon, a "nuclear pioneer" and "institution builder" by Dr. R. Chidambaram. Here was a man, who had interests and accomplishments that covered wide areas of nuclear science and technology; Western and Indian classical music; Indian philosophy; science policy, planning and administration; human resource development; and institution building – I was, in fact, speaking at a remarkable institution, a veritable think-tank that he had himself built. And yet, here he was, thinking so incisively about the possible depths of the macaque mind – I had hardly met anyone over the three years of my exploratory postdoctoral life so deeply interested in the macaques, whose cognitive capabilities had begun to intrigue me. Here then was an individual, who was as empathetic and full of wonder about these other beings, as I imagined myself to be. I realised finally I had found a home to come to.

And much to my delight, I soon received a call from Prof. Sreekantan asking me whether I would consider working in NIAS in the long term and if yes, whether I could come and see Dr. Ramanna at 9 am the following morning. I reached RR's office at 8.45 am the next day – it was a cool and fresh post-monsoon morning with blue skies, joyful cottony clouds and a chill in the air, so typical of Bangalore of those days – and waited nervously in another cane chair, just next to the column in front of his room. RR arrived at precisely 9 AM, stopped in his tracks when he noticed a visitor for him and looked quizzically at me; I think he could sense my anxiety.

RR: Hello, young man. Are you waiting for me? ... *a pause* ... Hey, aren't you the fellow who spoke about monkey consciousness the other day? What can I do for you?

I (very nervously): Yes, sir. Professor Sreekantan asked me to meet you in connection with the possibility of my working here in NIAS in the long term.

RR: Oh yes. So, what do you propose to do with the rest of your life?

I (slightly more confidently): I want to study the mind of the macaque, sir.

RR: Are you sure? You will never give up studying your monkeys, come what may?

I (definitely more confidently): No, sir.

RR (with a hint of a smile): Then you have a job. Go meet the Controller.

That was it. My first and last job interview ever. And I came home to roost, never to leave.

NIAS was an absolutely wonderful place to be in, especially in those days – and I do know it was RR who was able to imagine the space that would ultimately foster the kind of intellectual activity that he and J. R. D. Tata had envisaged for the institution. The buildings and the sylvan campus – which have been beautifully maintained till this day – immediately reminded me of what Martin Heidegger, the German philosopher, had said, in his famous essay of 1954, *Building, Dwelling, Thinking*:

"We do not dwell because we have built, but we have built because we dwell, that is because we are dwellers.... To build is in itself already to dwell.... Only if we are capable of dwelling, only then can we build".

I could almost imagine RR wanting to foster in all of us in NIAS a sense of belonging and connection to the environment, a call for a more responsible relationship with nature, recognising the importance of inhabiting this world in a mindful and respectful way. And in no way was this more evident than in RR's inordinate love for all beings, human and more-than-human, which I was to discover during my days with him.

RR would never cease to enquire about my macaques, whenever he met me, either in the quiet lecture hall – with cartoons by R. K. Laxman, which RR delighted in, on its walls – in the breezy corridors, at high tea in the late morning at the end of a Wednesday Talk – when Krishna or Satyamurthy (from the Institute cafeteria) would serve us hot *uddina vade* or *badanekayi bajji* – or in that little green patio, almost adjacent to his room, where we would enjoy our morning and evening tea and their ensuing conversations. RR's interest in my study subjects was, to begin with, academic – with often deep discussions of how one could define, detect or ultimately decipher consciousness in non-verbal beings, who could not, as he once remarked, tell us what they were thinking of. He was, of course, a true physical scientist, with faith in the laws, rules or forces regulating our existence on this planet, but, at the same time, he was surprisingly sensitive, very unlike several leading physicist friends of mine, to the unknown recesses of our mind and consciousness, with the firm belief that there were phenomena embedded in them that should be believed in scientifically for only then could they be explored by the principles and methodologies of the natural sciences. I even remember an incident, during one of our early institutional courses, when RR strongly argued in defence of scientific explorations of consciousness – both through biology and physics – against none other than the eminent neuroscientist and behavioural biologist, Prof. Obaid Siddiqi, who vociferously denied the very existence of consciousness as a definable phenomenon!

At the same time, RR was very concerned about the well-being of my macaques in a rapidly changing, increasingly human-dominated world. He had realised only too well that we were losing all our "natural" habitats to a rapidly urbanising world and on one occasion, even bemoaned the fact that ironically, the construction of NIAS, although destined to be a green haven of peace and tranquillity, had cost the campus of the Indian Institute of Science five acres of scrub jungle, typically native of the Mysore plateau, where the jackals still roamed. He possibly saw within me a spirit of concern for all the non-human beings that I was a companion and champion of, and this perception possibly never left him. It may have even brought me to NIAS in the first place!

The author with Con, an adult female bonnet macaque, one of his study subjects (Mudumalai Tiger Reserve, 2007, M. D. Madhusudan).

I remember another occasion in this connection, one that brought into focus RR's gentle sense of humour once again. On the late afternoon of 28 January 2000, we had all gathered in the little green patio to celebrate his 75th birthday, complete with a nice chocolate cake that he seemed to love so much. There was a group photograph ready to be taken and RR suddenly called out to M. D. Madhusudan, my and the institute's first doctoral student to graduate through NIAS' own doctoral programme, then established in collaboration with the Manipal Academy of Higher Education. "Hey you", RR exclaimed, his usually gentle voice raised slightly, "You come and stand next to me, the oldest and the youngest members of NIAS shall stand together". And then, suddenly having caught sight of me, he raised his voice even higher, "Hey you, you come and stand on my other side. You represent all the other species of this world!"

Other than his driving passion for music, both Western and Indian, and for his beloved piano, all of which remain well known, RR had a great love for the theatre. With his encouragement, NIAS came to be associated

Fig. 1.   Group photograph: Fifth NIAS Course on An Integrated Approach to Knowledge and Information for University and College Teachers; RR (seated fourth from left) gave a solo performance on the piano and delivered the valedictory address while the author (seated extreme right) was the course coordinator (NIAS, October 1997).

with the BLT or Bangalore Little Theatre, the oldest English theatre group in the city although this friendship was mainly propelled forward by our next Director, Prof. Roddam Narasimha – unimaginatively referred to as RN by us – and plays by various amateur groups came to be staged in our own J. R. D. Tata Auditorium, another brainchild of RR (Fig. 1).

But there was also a theatre movement born within the institution, once again vividly encouraged by RR and strongly supported later by RN. My close friend and then colleague at NIAS, Sundar Sarukkai – a philosopher of great repute but a rather sardonic individual – began to write short plays – usually on serious themes but laden with irony and sarcasm – and we began to perform them in our auditorium, usually with the enthusiastic support of both RR and RN. What, however, became more in/famous and even attracted audiences from outside, including members of the then NIAS Associates' Programme, were the little skits that Sundar, several members of the faculty, researchers or administrative staff and I put up, again rather regularly, often in the auditorium but also in informal spaces, such as the dining hall, and typically on the closing days of various conferences, schools or training programmes conducted by NIAS. These were fun performances, occasionally impromptu with *ad libbed* lines or scenes, but they revealed our inherent capacity to look beyond the seriousness of our regular work and daily activities in NIAS, and even more importantly, at the ability to laugh at oneself, at yourself, even though you may have been the architect of India's nuclear technology or of the country's aerospace programme, or even the founding fathers of a national policymaking institution that was NIAS.

Let me briefly describe one such skit, performed in the auditorium on one evening in January 1997, in front of an august audience, including RR and RN, attending the closing ceremony of NIAS' annual flagship course for senior executives of the country, one that continues even today. RN had just taken over as Director of NIAS, but he was still rather occupied handing over the reins of NAL, the National Aerospace Laboratories, of which he was the retiring Director, to his successor and hence, often absent from the institute. RR, on the other hand, seemed to be enjoying his new-found freedom from his various administrative tasks. He would take leisurely walks, with Chin-Chin leading the way and deciding the route to be taken by the two friends, along the extended narrow path, which runs even today between the then main building of the institute – currently its guest house and seat of its main administrative office – and the tree-laden parking area. The path would end, in those days, in a large green patch, which later became a medicinal plant garden, thanks to the efforts of Darshan Shankar and his non-governmental Foundation for the Revitalization of Local Health Traditions – currently the Trans-Disciplinary University – but which exists no more. A serious-looking Faculty Block of NIAS stands in its place.

In one scene of the skit, which was basically a tongue-in-cheek commentary on current affairs in NIAS, Sundar and I – playing ourselves, two relatively young members of the NIAS faculty – were discussing the two Directors of the institute.

Sundar: Hey Anindya, I am a new faculty in NIAS. Wasn't Dr. Raja Ramanna its first Director?

I (seriously): No.

Sundar: Really? But that's what I have heard. Then who was it?

I (very sternly): Chin-Chin.

Sundar: Oh! And what was Dr. Raja Ramanna's role in the institute then?

I: Personal Assistant to the Director of NIAS.

Sundar (relieved): Ah, now I see why NIAS has gone to the dogs. And who is the current Director of NIAS?

I (smilingly): We don't have one.

Sundar: Oh!

I: Well, we apparently have a director, but I think it's an urban myth. No one has apparently ever seen him. He has never been here.

Sundar: What does that mean?

I: It's all aerospace out here. After going to the dogs, we seem to be just disappearing into thin air!

I still remember the laughter and cheers that greeted this irreverent depiction of NIAS, the loudest clapping from the two Directors, who sat in the first row of the audience. I also still remember how, when we came down from the stage, RR, as was customary of him, wrapped his arms around Sundar and me and invited us to join him in a guest suite for some rum – Old Monk it was – which we were all so fond of, to celebrate, as had also become customary, the successful end of yet another senior executives' course. All was well.

Let me now end this rambling reflection of the twilight years of RR's life and times in NIAS with a last recollection. All of us would regularly have lunch together in what was then the dining room of the institute – now a rather dilapidated students' common room – and occasionally light-hearted conversations would flow, sometimes over the chicken curry, cooked delectably by Krishna and served at least twice or thrice a week, days that I personally truly looked forward to. There were three tables placed in parallel across the relatively small room – yes, all of us in NIAS then would comfortably fit in with even half a table to spare; RR would customarily sit at the head of the far end of the first table, facing the door through which we would file into the room. He would greet each one of us, as we entered, with a gentle smile or a slight nod of his head – the benevolent head of this small family.

In May 1998, India conducted a series of nuclear tests, codenamed Operation Shakti, at the Pokhran Test Range in Rajasthan. There were intensive discussions among us in NIAS, some at lunchtime, with divided opinions about the future goals of India's nuclear programme flying fast and furious across the tables. On one of these occasions, RR suddenly looked at me, sitting at the second table, and loudly said, "Hey you, I hear you are against the bomb?" I smiled and quietly nodded, to which his immediate riposte was, "How did you get this job at NIAS?" There was loud laughter across the room but no tension of any kind whatsoever. At that time, RR had been advocating for strict policies to prevent nuclear proliferation globally and had even argued against further nuclear testing in our country. But his almost Buddhist ability to hold different view-points within his own mind, to argue them out rationally and to reflect on them, without any regret over any decision made in the past, all clothed in his gentle sense of humour, remained with him till the end. All was always well.

Finally, let me assure you, RR, that I am still with my macaques; I have kept my promise to you. And even now, I seem to hear the strains of your piano emanating from the auditorium, gently playing the closing bars of the first movement, *Adagio sostenuto*, of Beethoven's *Sonata quasi una fantasia*, which you loved so much. I can also see you quietly walking down the sylvan path of your beloved institute, your faithful companion Chin-Chin still gently trotting alongside, as the evening stars shine down on both of you.

Anindya Sinha, a Professor at the National Institute of Advanced Studies in Bangalore, had early research interests in the molecular biochemistry of yeast metabolism, the social biology of wasps and the classical genetics of human disease although his principal research, over the last three decades, has been on the behavioural ecology, cognitive ethology, population and behavioural genetics, evolutionary biology and conservation studies of non-human primates. His current research in the natural philosophies, urban ecologies, art heritage and performance studies primarily concern ethnographic explorations of human–non-human relations and the lived experiences of non/humans, with their promise of unique understandings of more-than-human lifeworlds, in the past, today and in the future.

Chapter 23

A Daughter Remembers

Nina Kanjirath

B1, Vishwa Apartment, 12/9 Lavelle Road, Bengaluru 560001

ninakan25@gmail.com

In the early hours of the morning of 22 June 1982, Air India Flight 403 from Singapore was approaching Bombay airport. Most passengers were fast asleep. The heavy monsoon rains had reduced visibility to near zero, thunderstorms raged overhead and the airport was flooded. Given the dangerous conditions, the pilots had been asked to divert the aircraft to Madras (now Chennai), but for reasons known or unknown, they continued towards Bombay and lowered the landing gear to prepare for landing as they approached the runway.

Since the authorities were not expecting them, the landing lights had not been switched on. On realising this, the pilots tried to take off and flew upwards for several seconds. The plane was unable to sustain this move and came crashing down without warning at Santa Cruz Airport, bouncing violently on the tarmac before skidding off the runway.

Among the handful of passengers in the first-class cabin was my father. Startled passengers woke up to what they thought was a very very rough landing. The aircraft had veered off the tarmac and come to a halt in a soggy, putrid bog next to Dharavi (one of Asia's largest slums).

Miraculously, the plane did not catch fire, possibly because of the wet, muddy terrain that absorbed much of the impact. Inside, everything was eerily silent. The power had failed, leaving passengers in pitch darkness, with only the deafening sound of rain hammering against the fuselage.

In the first-class section (which had broken off from the body of the plane), about half a dozen passengers, including an army officer, waited for the crew in the dark but there was no response. Finally, the officer managed to locate and wrestle open an exit door. The small group, including my father, slithered down into the bog and moved onto the tarmac. In the darkness, they could see lights at a distance and assumed that it must be the airport. With no other option, they began their long-wet walk through the relentless rain and flooded tarmac towards those lights.

When they finally reached the airport building, they were relieved to see a security guard standing by a door and rushed up to him.

"We have just come off the plane", they said, "and we want to go inside". The guard shook his head and said "I can't allow you to enter because this is a door for exiting passengers only. The door for entry is on the other side of the airport".

Unaware of the seriousness of the situation, they trudged around the building until they found the main entrance. By this point, my father was exhausted and decided to go home. He bid farewell to his fellow passengers and made his way towards the parking lot, hoping to find his car.

As he approached his car, he saw his secretary and a few others standing by his vehicle, their faces bowed, grim and downcast. They had heard about the crash and believed him to be among the dead. When they looked up and saw him bedraggled, covered in mud and very much alive, they froze in shock. For a moment, they thought they had seen a ghost.

Without much ado, my father got into his car and went home.

When he arrived, my mother asked why he was late. "There was a rough landing", he said, "There was no bus and we had to walk all the way from the plane to the airport". He was exhausted, his body protesting from the sheer sudden drop of the plane and he went off to sleep.

An hour later, the phone began ringing continuously. Friends, relatives and colleagues were calling in panic asking my mother (who learnt about the crash from the callers) if he was alive and alright.

News had spread like wildfire. Air India Flight 403 had crashed, broken into three pieces and 17 people had lost their lives and several were missing.

In the aftermath, speculation swirled as to why the plane had crashed, and it became a mystery. In the meantime, there were rumours that there could have been a bomb planted on the plane to target my father who had led India's nuclear programme.

At the time, I was recently married and had been working with Air India and was hoping to take advantage of AI's policy that permitted employees to take a year-long sabbatical, as I wanted to join my husband in Colombo with our newborn daughter. This offer of leave had been in motion for many years till 1982 when it was decided to withdraw this offer.

In the days that followed, many senior airline officials visited my father every day to express their deep regret and apologies for the ordeal he had endured. They asked if there was anything they could do for him. He suddenly recalled my request for a sabbatical and asked if that leave would be possible.

At first, I had no idea about my father's intervention. But soon, I noticed a shift when some of my colleagues' attitudes changed towards me.

I realised I had been privileged, so I asked my father to withdraw the request, He said "Look, they have been asking if they can do anything for me – make amends for this accident, so I made a request. If they cannot do it, let them tell us that it's not possible and we will accept the outcome".

Then, there was an inquiry as to why the plane landed in Bombay.

This was the early 1980s before PM Manmohan opened the economy, and the flight was coming in from Singapore.

It crashed very early in the morning next to the Dharavi slum area.

There were questions on what exactly had been in the plane's cargo hold and speculation that it may have contained contraband to be cleared at Bombay airport. But by the time the officials got round to collecting any proof of this, the plane's scattered cargo contents, all the suitcases and personal belongings of the passengers that were strewn all over the tarmac and Dharavi were gone!

Nina aka Rukmini Kanjirath's background has been in education since 1983. She has taught in several international schools in India and overseas and in schools for underprivileged children. She designed and set up Gaia Preschool in Bangalore which is attended by several of her students' children and now works as a teacher trainer. She is passionate about theatre and all genres of music and particularly enjoys the creative aspect of design in all spheres.

Chapter 24

Prof. Raja Ramanna: Enigma Viewed from a Kaleidoscope

Malavika Kapur

National Institute of Advanced Studies, Bengaluru, India

malavikakapur@gmail.com

Preamble

A confession! When I was asked to contribute a chapter to the Raja Ramanna Centenary Celebratory Volume, I was most surprised as well as delighted. Surprised as I hardly knew him in person. Delighted as I knew of him having a ringside view of him as a faculty wife along with others in Prof. Raja Ramanna's charmed circle at the time of the inception of the National Institute of Advanced Studies at the Indian Institute of Science Campus in Bengaluru. These were Profs. M. N. Srinivas, an eminent sociologist, B. V. Sreekantan, a renowned scientist, and C. V. Sundaram, a close confidant of Prof. Raja Ramanna. These were the gems belonging to different disciplines brought together and bound by great camaraderie; so were the wives of the gracious founders, except for C. V. Sundaram who was a bachelor. We were small families and got to know each other well over time.

Before Prof. Raja Ramanna got to know my husband, Dr. R. L. Kapur, both of us were on the faculty of the National Institute of Mental Health and Neurosciences, Bengaluru, having returned from Edinburgh, Scotland, after his study with "Great Universe of Kota" in South Kanara.

I worked as a faculty in the Department of Clinical Psychology and Dr. R. L. Kapur as the Professor of Community Psychiatry. I decided that I would put my academic career on hold, to take care of the schooling of our two children, while my husband networked in India and abroad through his World Health Organisation (WHO) and Harvard connections.

There was no way I could get to know about Prof. Raja Ramanna. In this kaleidoscopic account, I have brought together what I saw in fragments into a complete picture of the complex and enigmatic man Prof. Raja Ramanna was, across the span of years, since NIAS came into existence.

RLK (referred to Dr. R. L. Kapur in the Mysore tradition) had three mentors and three gurus. Prof. Raja Ramanna, Dr. R. Varma and Prof. George Morrison Carstairs were his mentors. Drs. J. S. Neki, N. C. Surya and Ernal Hoch were his gurus. This is a homage to the most exceptional and best of them in kindness and affection. He was the only one who did not get riled by him despite provocation!

Beginning of National Institute of Advanced Studies (NIAS)

The dream and vision of Mr. J. R. D. Tata were handed over to Prof. Raja Ramanna to bring to life. Prof. Raja Ramanna and RLK were the two who were capable of braiding the disparate bands together and bringing them alive. So, he was introduced to the charmed circle of JRD Tata and Homi Bhabha. At the time of the planning of NIAS, Prof. Raja Ramanna was a scientist extraordinaire of nuclear energy. For obvious reasons at this juncture, he was made the Defense Minister of India, an ideal blend of a scientist and diplomat. Only he could possibly hold that critical position at a critical time in the political future of the country. He deputed RLK as the Deputy Director of NIAS at that juncture (Fig. 1).

The thread that bound the two together in the overspecialised world of various sciences and humanities was dreamed of by Jamshedji Tata who created the Indian Institute of Science. J. R. D. Tata was bent upon to make it into reality by creating the National Institute of Advanced Studies. On the other side of the city of Bengaluru Dr. M. V. Govindaswamy, created the All-India Institute of Mental Health adopting a first-ever multidisciplinary approach to service, research and training in all disciplines related to mind and brain, including Indian psychology, RLK and I were

Fig. 1. (Left to right) Dr. R. L. Kapur, Dr. Raja Ramanna, one of the participants and the author during the Seventh Course for Senior Executives at NIAS (1993).

in the early batches of post-graduates sharing that vision of multi-disciplinarity. Is it a wonder, that Prof. Raja Ramanna's and RLK's visions and mission melded in the creation of the National Institute of Advanced Studies?

The following are some instances of how the two related to each other. Their personalities were diametrically opposed to each other but there was a strong affectional bond between the two. The following instances are trivial but lead to deeper insight into their relationship.

RLK never considered anyone superior in status, who needed to be flattered or worshipped. He dealt with those above him in status as equals to him in his interactions. One such instance was when he admonished Prof. Raja Ramanna for overspending on a hospitality bill saying that he had spent too much NIAS money! Prof. Raja Ramanna sheepishly smiled! That's all! Another time, Prof. Raja Ramanna had just returned to his NIAS office following major cardiac surgery in Mumbai. He had heard that we had new Alsatian puppies that we were giving away to friends. Prof. Raja Ramanna wanted a puppy. RLK refused point blank saying "You just have had a major surgery; you cannot take care of a boisterous large dog! I will not give a puppy to you!" I am sure he was disappointed. But a few days later, a tiny Lhasa Apso accompanied him to his office to become a constant presence. He had chosen a special name for it (Chin-Chin). Prof. Raja Ramanna got his sweet revenge most nicely.

Some Sparkles at NIAS

To me, the Wednesday Discussion Meeting represented Prof. Raja Ramanna's whole persona, something for which he spent considerable time and thought.

Usually, excellent speakers from within NIAS or outside were invited to speak for an hour. This was followed by tea and tiffin in the dining hall. There were reserved tables for outsiders. The rest of us, postgraduates and faculty, joined in. Prof. Raja Ramanna sat in his usual place.

The previous day, he would spend a long time with the chef about marketing seasonal vegetables, what to prepare for Wednesday and how to prepare it.

An example would be "Avare Kalu" (hyacinth beans, a common ingredient in kitchens in Bengaluru during the winter and spring) season. There were several recipes to be concocted, bondas and chutneys to be made!

Extraordinary diligence was required by the kitchen staff! Then, when it was appreciated by everyone, he would be satisfied.

As the talk got over and the participants discussed with the speaker, Prof. Ramanna took his chair, and someone among us at a hearing distance would offer his uproarious one-liners on the talk. These were the best and briefest views of the presentations.

The people stayed on and discussed for a long time, so the lunch was delayed.

These were Prof. Raja Ramanna's soiree or saloon moments which we never missed. In one instant, he would demolish the argument laid out by the speaker. Even genuine appreciation came out cryptically. But truth must prevail. He always spoke gently with an enigmatic smile!

Prof. Raja Ramanna and Music

Western classical and South Indian classical music were his passions. He was an accomplished pianist and so was his younger daughter Nirupa. On occasions, at home or in Institute get-togethers, they would both play the piano together. When Nirupa accompanied him, she always played well, his expression was beatific and paternal pride was something to be seen to be believed. It was not the vain pride of a brilliant father but it was the father's love sharing something they both loved. Sharing something precious with a child is something I witnessed among some exceptional fathers.

Encounters with Antagonists

I would like to share an event I found most amusing. My father, Shivarama Karanth, was a doyen of Kannada literature but also a firebrand antinuclear activist. He had gone to the Supreme Court against the Kaiga Power Plant and lost the case. Prof. Ramanna told me "I want to invite your father for a talk at NIAS. But I am scared of him!" I too thought it was a bad idea, though I am a believer in my father's ideology. So, the encounter never occurred much to my delight!

Lighter Moments

Prof. Ramanna was a great company to Dr. Kapur of shared happy hours, whether in Bengaluru, Mumbai or Delhi, both loved their scotch together. However, the special concoction developed by Prof. Ramanna was Old Monk with green chillies, which was not Dr. Kapur's favourite. This bonding with Prof. Ramanna was very special for Dr. Kapur as he would never drink in the presence of his father and my father, out of respect. Such was Dr. Kapur's bond with Prof. Raja Ramanna.

The Affectional Bond

Prof. Raja Ramanna was a man of exceptional humility and integrity. Dr. Kapur was a deep scholar in multiple disciplines as well as a physician, psychiatrist and extraordinary healer. His relationship with Prof. Ramanna's family was very close and Prof. Ramanna was truly a family man. Obviously, the relationship of Dr. Kapur with the various members of the family had to be very complex and nuanced. But Prof. Ramanna's life at home was not a bed of roses. His home life was harder than the outside world he had negotiated so successfully. Here was where Dr. Kapur's support was needed and unstintingly provided by him. He had worked hard to reduce his burden on the family.

The bond between the two was deep and strong. Dr. Kapur's own father was authoritarian and confrontative, while Prof. Ramanna was a loving tolerant father.

In comparison to luminary counterparts with whom Dr. R.L. Kapur had similar demanding relationships, Prof. Raja Ramanna stood above head and shoulders in his personal integrity.

Conclusion

Prof. Raja Ramanna melded science and philosophy, wisdom and humour, trust and empathy and navigated his life with a moral compass. He was an enigma. He was a warrior. What better name to adorn him than Savyasachi? The imagery came to me when I recently listened to Dr. Ramamurthy evocative lecture on the secret path to the success of the Pokhran test.

Prof. Malavika Kapoor is an Honorary Professor at the National Institute of Advanced Studies, Bangalore. Until recently she was the Professor and Head of the Department of Clinical Psychology at the National Institute of Mental Health and Neurosciences, Bangalore. She has 7 books and over 100 papers to her credit. A Fellow of the Indian Association of Clinical Psychologists, the Indian Association of Child and Adolescent Mental Health, the National Academy of Psychology and the British Psychological Society, she has also been a consultant for the World Health Organization and the India Council for Medical Research among other coveted organisations.

She is also involved in the development of assessment tools and intervention packages for children and adolescents in the Indian context. Her main contribution is her work of developing integrated models of mental health service delivery for children and adolescents.

Chapter 25

My Brief Interactions About the Book *The Structure of Music in Raga and Western Systems* by Raja Ramanna

M. B. Rajani

National Institute of Advanced Studies, Bengaluru, India

mbrajani@nias.res.in

It is an immense honour and privilege to be part of this volume that is being brought out in memory of Dr. Raja Ramanna on his birth centenary. Though I am relatively young compared to the active years of Dr. Ramanna's illustrious career, I am proud that my time at the National Institute of Advanced Studies (NIAS) overlapped with the last six months of his time here. I vividly recall several occasions when we met during those six months. One particular memory that stands out is from April 2004, when I gave my first Wednesday Talk at NIAS on my master's research, Dr. Ramanna attended the talk, and afterwards, he spoke to me personally about the talk. I remember how he would go to the dining hall early, around 12:30, and I would time my arrival to catch a moment to greet him. On the third day of doing this, he chuckled and remarked, "It amazes me how young people can feel hungry so soon!"

While he was a luminary, a world-renowned scientist, whose deep and passionate interest in music was well known, the genuine interest and engagement he displayed when over a casual lunch table discussion, he discovered my interest in the theory of Carnatic music, was truly inspiring. He was interested to know that I had studied musicology under the tutelage of Dr. John R. Marr (a British Indologist and an expert in South Indian musicology). He has generously acknowledged Dr. Marr's contribution in the preparation of his book *The Structure of Music in Raga and Western Systems* (Ramanna, 1993), published by Bharatiya Vidhya Bhavan, Bombay.

This book is a thoughtful exploration of music theory and structure in both Indian and Western classical traditions. Ramanna, a prominent physicist and musician, approaches this comparative study with a rare blend of scientific precision and artistic insight, offering readers a unique perspective on the underlying principles that shape these two musical worlds.

The book is divided into sections that methodically explore the conceptual foundations of Indian raga and Western tonal music, focusing on scales, melody, harmony, rhythm and improvisation. Ramanna delves deeply into the raga system, examining its melodic foundations, its emotional impact and its inherent flexibility that allows for extensive improvisation within specific structural limits. By contrast, he presents Western classical music as being rooted in harmony and structured forms, which prioritise the vertical stacking of sounds to create complex tonal relationships. Ramanna's background in physics adds a layer of depth, as he explains how mathematical relationships play a crucial role in shaping scales and intervals in both systems.

One of the strengths of the book is its accessible writing style, which makes complex music theory concepts understandable for both musicians and general readers interested in music. Ramanna employs diagrams and examples to clarify these ideas, making the comparisons between raga and Western systems more relatable. While he does not heavily favour one system over the other, he illuminates their unique strengths: the emotive, improvisational freedom in raga and the structural complexity of Western harmony.

The book deeply engages with the Melakarta system, which provides a theoretical foundation for Carnatic music. The Melakarta system was conceptualised in the 17th century by musicologist Venkatamakhi. The standardised system in use today is said to have been formulated by Govindacharya in the 18th century. The Melakarta system is a foundational framework in Carnatic music that organises ragas into a systematic structure based on a 22-note scale (or 12 semi-tone in one octave), which includes microtonal variations of the seven basic notes (svara): Sa (Shadja), Ri (Rishabha) with three variations (R_1, R_2, R_3), Ga (Gandhara) with three variations (G_1, G_2, G_3), Ma (Madhyama) with two variations

Table 1. Notes of Melakarta system, a foundational framework of Carnatic music.

Semi tones	1	2	3	4	5	6	7	8	9	10	11	12
Note	S	R_1	R_2	R_3	G_3	M_1	M_2	P	Dh_1	Dh_2	Dh_3	N_3
Overlapping notes			G_1	G_2						N_1	N_2	

(M_1, M_2), Pa (Panchama), Dha (Dhaivata) with three variations (Dh_1, Dh_2, Dh_3) and Ni (Nishada) with three variations (N_1, N_2, N_3); some of them overlap with other, as illustrated in Table 1. It consists of 72 parent ragas, known as Melakarta ragas, which serve as sources for countless derived or janya ragas.

Each of the 72 parents or Melakarta ragas has one each of the seven notes (Sa, Ri, Ga, Ma, Pa, Dha, Ni); the variations (in the Ri, Ga, Ma, Dha and Ni) allow us to create various combinations adding up to 72. The Melakarta ragas are symmetrical, that is, the same seven *svaras* in both ascending (Arohana) and descending (Avarohana) sequences.

This existing systemic structure in Carnatic music provided a basis for exploring the similarities and differences between the structure of Western classical and Carnatic music which Ramanna's book has explored very systematically. In the book, apart from providing staff notation for several ragas (parent, child and derivatives), tables and graphs are extensively used to organise material and display structure. Here is one example where the connection between six of the Melakarta ragas is displayed in the form of a circular graph, which also shows the corresponding scale in the Western system (Fig. 1).

Our discussions on music theory spurred on a casual lunch table discussion when I shared with him about my attempt to capture the Melakarta system in the form of a concentric double circular diagram, where the top circle would rotate revealing the details of each of the 72 Melakarta (Fig. 2). This interaction has remained a very memorable one. Unfortunately, we did not have more time together to work on the same.

Dr. Ramanna expressed a desire to revise his book for a second edition and asked me to coordinate a meeting with my mentor, Dr. John Marr, who was to visit Bengaluru shortly thereafter. Our last conversation was on his final day at NIAS when he inquired about the dates of Dr. Marr's next visit. He also mentioned that once he returned from Bombay, he would give me an updated copy of his music book to pass on to Dr. Marr.

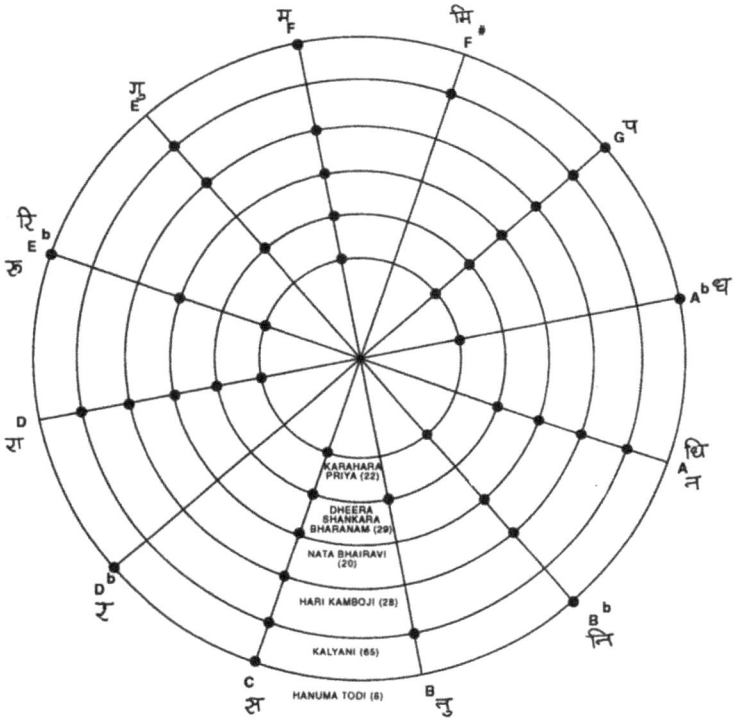

THE FORMATION OF RAGAS MECCHA KALYANI (65)   HARI KAMBHOJI (28)
KARAHARA PRIYA (22)   SHANKARABHARANA M (29)   AND NATA BHAIRAVI (20)
BY TONIC SHIFT FROM HANUMAN TODI (8)

SQ. SIGNATURE 212212
FIG. 3 (f)

Fig. 1.   From *The Structure of Music in Raga and Western Systems* by Raja Ramanna, p. 50, 1993.

The Structure of Music in Raga and Western Systems is an enlightening read that offers a rare and balanced view of two rich musical traditions, inviting readers to appreciate the diverse paths that music can take to express universal emotions and ideas.

Although I am not a recognised expert in the field, musicology has always been a serious parallel interest of mine. I am grateful for the

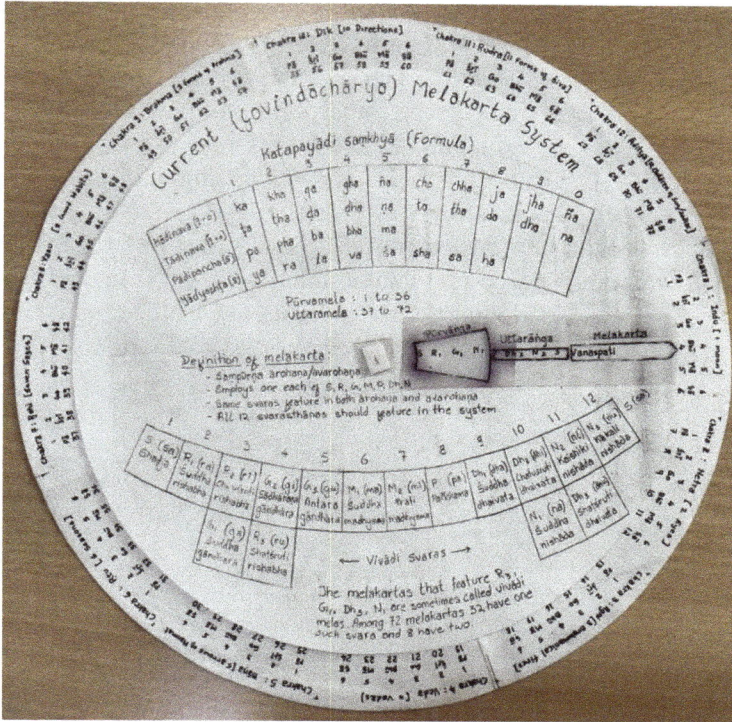

Fig. 2. Melakarta wheel prepared by the author (2003).

opportunity to contribute to this volume, as it not only allowed me to reflect on Dr. Ramanna's work but also gave me a chance to imagine the conversations we might have had if he had returned from Bombay.

Dr. M. B. Rajani is an Associate Professor at the National Institute of Advanced Studies (NIAS), Bangalore. Her primary research has two inter-related facets: analysing cultural landscapes using geospatial data to identify new features of archaeological interest and advancing the usage of such analysis towards preservation of built heritage in the face of rapid urbanisation. Her other research interests include inter-cultural exchanges during colonial period in visual arts, music and cartography and musicology of classical Indian music.

Chapter 26

The Alchemy of Mysuru, Music and the Maverick Genius of Dr. Raja Ramanna

Deepti Navaratna

National Institute of Advanced Studies, Bengaluru, India

deepti.navaratna@nias.res.in

The Enlightenment and Romanticism were two ideologically opposed movements that shaped the intelligentsia of the 18th century. While Enlightenment placed an impetus on science, knowledge, reason, method, objectivity, logic and evidence, Romanticism highlighted the significance of art, experientialism, imagination, spontaneity, individuality, creativity and empathy. Although dislocated from these movements in terms of time and space, fast forward to a century later, one finds beautiful synergies to both these approaches to life in Mysore. Princely Mysore forged a syncretic culture where the multiplicity of such pursuits was seen as desirable. Leading by example were the Wadiyar kings themselves, who in addition to being adroit statesmen were scholars, musicians and philosophers in a long line of pedigree monarchy. The Maharaja of Mysore, His Highness Jayachamarajendra Wadiyar, who was piano guru-bandhu to Dr. Raja Ramanna cautions against the stereotyping of Eastern spirituality and Western materiality: *"It has become far too common to describe the two predominant viewpoints of the world as Western and Eastern. It has become the vogue these days, in keeping with the mania for labelling everything, to label the Western view as materialistic and the Eastern as spiritual. This is to rest on a false foundation. Perhaps, we may admit that there is much of truth in it, that by and large, the West has been more concerned with outside things pertaining to the world of matter more than with the inner things of the spirit. Equally well can the charge be sustained against the East that it became obsessed so much with the inner side of life that it forgot the existence of an external material world. Therefore, such generalizations do no one any good – least of all to the protagonists of such views".*

Princely Mysore in the 21st century, in addition to being a model state, also boasted of a culture which looked at the East-meets-West encounters as an opportunity rather than as a liability of the current. Dr. Ramanna often credits his renaissance mindset to a Mysorean lifestyle: "*The ancient city of Mysore with its gentle and courteous people moulded by the traditions of the court, remains a place of beauty and hope. Mysore is a refuge*". The idea that human genius can be as at home with science as with music or the arts. The audacity to demand more from the human potential, to explore new possibilities of existence and to chart new sensibilities. To value tradition and the ancient while embracing the joy of the exotic or to explore a new aesthetic. Such pursuits sat well in Mysuru, which managed to curate many brilliant minds thriving in an alchemy of their colonial realities and their deep connections to Indian knowledge systems. The beauty of this spirit of his generation of Mysoreans was that they never problematised these polarities; they subverted socio-political realities to forge a brave new world for themselves. The products of such a spirit also speak for themselves; case and point, Dr. Ramanna's book, authored in 1993, *The Structure of Music in Raga and Western Systems.* How else can one cope with the beautiful reality that the man who became the much-touted face of India's nuclear programme also held a Diploma from the Royal School of Music, London! As a faculty of NIAS, and as someone who loves science as much as music, in his biographic accounts, I find many lessons and synergies. The world loves its boxes and disciplines, rightfully so. Every so often comes a disruptive genius like Dr. Ramanna – fully defying such outer boundaries to redraw an interdisciplinary map – where paramatma and paramanu meet. To make this unboxed zone your kingdom takes a Raja indeed!

Dr. Ramanna was the great-grandson of Bindiganavale Venkatappa Krishna Iyengar, the first Indian and Kannadiga to serve as Deputy commissioner in the princely state of Mysore. His family lineage is well known in Mysore circles; his father was in the Mysore judicial service – it was indeed a life of great privilege shaped at the plexus of great opportunity. *"In my early life, it was my mother Rukminiammal, who really taught me to appreciate our arts. An old British missionary, Margaret Moffet, developed my interest in piano; she even taught me how to eat with fork and knife"*. Education, for him, seemed to run in two parallel heterophonies – a melody that placed him as part of a continuous chain of Indian knowledge and another which looked out of this psyche with a modern, scientific pursuit of Western education. *"A fantastic storyteller, Rajamma would often tell me stories from the Puranas and the great epics. In retrospect that was the best education I ever received. I'm proud of the fact that Raja, the name by which I am referred to by all my friends, is taken from my aunt's name – Rajamma"*. Some of the major themes of his life – scientific enquiry, musicology and Adhyatmic discourse – could be traced back to his early childhood immersions in Mysore. *"Growing up in the British days is one of the happiest memories. Bangalore, where I grew up, was an old cantonment and the British way of life early on had a strong impact on me. The unhappiest memory is when I was six or seven years old – my older sister died of typhoid. She encouraged me to take up European music"*. Sister Doraichetty, who learnt piano with Dr. Ramanna in his early years at the Good Shepherd School in Mysore, recollects how he would often come back to school to play on the very same piano on which he was taught as a child. *"I was shifted to the Good Shepherd Convent which was located on the outskirts of Bangalore. The nuns of this convent had taught the members of the royal family and enjoyed a good reputation. Apart from that, the main advantage at this school was that they also taught European music"*. As reported in some of his informal discussions, Carnatic music soirees were a staple at his house, thanks to the family's long-standing ties with the art of culture societies in Mysore. Some of the best Carnatic musicians, literary figures and Hindustani musicians of that time would indulge in intellectual discussions on the intersections of culture, aesthetics and philosophy. Dr. Ramanna recollects that *"at home, there was now the general feeling that because there was enough appreciation of Carnatic music somebody should also study European music. It was decided eventually that I make the effort and so began my piano lessons at the new school at the ripe age of six. I guess the nuns at the convent must have been conscientious, but I was not particularly attracted to any of them except for one outstanding lady, an Irish nun called Mother Maurice. She had been the music teacher to the Yuvaraja's son, Jaya Chamaraja, and all the princess of Mysore court"*. Thanks to Mother Maurice's loving attention, a whole generation of pianists were rising to public popularity. *"The Maharaja also patronized a host of Carnatic and Hindustani musicians, as was the tradition of the time. Word reached him, through various sources, that I could play the piano well and an audition was fixed for me at the Jaganmohan Palace in 1937"*. One finds that music played scriptwriter and director at crucial scenes of his operatic life. Music cast him in his best roles and amid the most august audiences, with great ease. It was music and Mysore again that played conductor to his next big landmark later in his life – his meeting with Dr. Homi Bhabha.

Music, in his life, seemed to make all life transitions smoother – like a kind witness. Recollecting his family's move to Bangalore, Dr. Ramanna writes about adapting to the famous Bishop Cottons School in Bangalore, known for its stiff British education and discipline: *"Although I managed to do well in school as far as studies were concerned, I still felt somehow a misfit as I couldn't conform to a major activity in the curriculum set up by the British sports. However, that did not pose a great problem because I'd another support system–music"*. This anchoring influence of music remained strong, for, when asked to describe his perfect day, at the peak of his Smiling Buddha project, he said, *"Playing the piano by myself, with no one around. One derives a great sense of self-happiness"*. Music also brought into the loving countenance of the then Maharaja Krishnaraja Wadiyar IV, *"On the day of the audition, the Maharaja listened intently to a new set of pieces that I played for him. Later, he came up for a chat and asked whether my teachers were guiding me properly and whether they discriminated between me and the European children. I was touched. The Maharaja was genuine in the care he showed towards a 12-year-old"*. While at the Madras Christian College, his vacations back home were steeped in music – catching up with the Mysore music "scene". *"My vacations from Tambaram were spent in Mysore and I resumed my association with the Mysore Palace band and music. Dr. Mistowski, my teacher, one day said, 'a famous Indian Scientist is here, you must know him, his name is Homi Bhabha'! My meeting with Bhabha would determine the course of next several years of my life"*, he reminisces in his biography *Years of Pilgrimage*. What transpired after this meeting in 1944 is history and changed the trajectory of his life.

The Mysore English and Carnatic orchestras were very dear to his heart – many of the band members and retired orchestra musicians fondly remember Dr. Ramanna playing and rehearsing with them. As early as 1868, the Palace Bands in Mysore were established in line with the philharmonic idea of looking at a variety of world music as inspiration for European art music. Far far away from London, which was the epicentre of such thoughts in England, Mysore adopted this philharmonic spirit by birthing the music bands. The English Band playing Western music and the Carnatic Orchestra playing Carnatic classical music on exotic instruments were created as two parallel lives animating a single spirit. The music bands were to change the musical ecosystem of Mysore and became an integral part of the social fabric of Mysore in the years to follow. They became a staple in Dr. Ramanna's life; he often performed with them – flying back to his musical nest of sorts in Mysore. What is fascinating in his story is that throughout his life, he never let go of what soothed his soul, as he adventured through great new beginnings, difficult and high-stake projects and perhaps some not-so-sweet moments that have populated his professional career. While studying for a B.Sc. degree in Physics from the Madras Christian College, his musical avatar also got a B.A. degree in Classical Music. While studying for a Ph.D. in Nuclear Physics at King's College in London, he obtained a *Licentiate of the Royal Schools of Music* degree. Music and science found a resonant alchemy in his world, side-by-side, like the relationship between the vadi and samavadi notes in a raaga, like the tonic and the dominant in European art music.

Dr. B. V. Sreekantan, one of the founding members of the National Institute of Advanced Studies and a colleague of Ramanna at the Tata Institute of Fundamental Research, has documented some episodes of the TIFR: "When Ramanna joined TIFR, the institute had just been shifted from its first premises at Kenilworth, 54, Pedder Road, Cumbala Hills in Bombay to the Yatcht Club premises and alteration work of the building was in full swing. The so-called servants' quarters of the Yacht Club were converted as the hostel for unmarried scientists of TIFR. Bhabha, who had known Ramanna's interests and abilities in music, allotted him two adjacent rooms in the top-most fourth floor of the hostel, one for Ramanna and the other for his piano. The investigations carried out by Ramanna and his co-workers on light charged particle emission in fission induced by thermal and fast neutrons provided important insight on the mechanism of emission of these particles. The stochastic theory of fragment mass and charge distributions in fission is a unique contribution of Ramanna to fission theory".

Henry David Thoreau, a famous American naturalist and philosopher, waxes romantic: "*When I hear music, I fear no danger. I am invulnerable.*

I see no foe. I am related to the earliest times, and to the latest". Perhaps, it would be no hyperbole to say that in Dr. Ramanna's life, music made his inner Buddha smile. The romance between science and music – as lived out in the life of Dr. Ramanna – is almost idyllic, and yet so real, serving as a perpetual inspiration for generations to come.

Dr. Deepti Navaratna is a musician and neuroscientist interested in the alchemy of research and performance. She is currently the T. V. Raman Pai Chair Professor at the National Institute of Advanced Studies, where she heads the Music, Brain and Creativity Laboratory – dedicated to research at the intersection of neuro-education, musicology and creativity. After several years of neuroscience research at Harvard Medical School, USA, her current research interests span neuroscience, arts and humanities, exploring new frontiers of interdisciplinary knowledge, such as cognitive hermeneutics, neuropsychology and empirical musicology. She is a Chevening Clore Fellow (2021–2023), recognised by the UK Government as one of the most creative artists and dynamic leaders across the globe. In 2021, she was invited to perform at the Parliament of World Religions in Chicago, where she premiered her inter-faith music diplomacy project "Dialogues with The Divine".

Until 2022, she served as the Regional Director, Indira Gandhi National Centre for the Arts, Ministry of Culture, Government of India, Bengaluru, India. She directed the Centre's research, academic and outreach activities at the intersection of science and humanities – with a focus on Indian knowledge systems. An accomplished South Indian Classical musician, her music has been presented at premiere performing spaces in the United States, such as Asia Society (New York City), Symphony Space (New York City), Jordan Hall (Boston), Harvard Arts Museum (Cambridge), Museum of Fine Arts (Boston), Yale School of music (New Haven, CT), Indian Council for Cultural Relations (ICCR) Sri Lanka, Central Conservatory, Beijing, and Tsing Hua University, Taiwan, among others. She has received several distinctions as a traditional musician such as the India Foundation for the Arts - Research Grant (2016), Cambridge Arts Council Education and Access Grant (2011 and 2015), Emerging Artist Award from St. Botolph Foundation (2011) and the Traditional and Ethnic Arts Fellowship from the Utah Arts Council (2009), Cambridge Arts Council Fellowship (2014), among others. Her foray into the world of history and letters was through The Maverick Maharaja – the first definitive biography of the Maharaja of Mysore, Jayachamarajendra Wadiyar, in 2022.

Chapter 27

A Tryst in Music, Half A Century Apart

Anant Kamath

National Institute of Advanced Studies, Bengaluru, India

anant.kamath@nias.res.in

All of us thought this would be his swansong. Only a few of us could actually believe it was the man himself, in the flesh, at the piano, but very evidently his fingers did not cooperate with his mind. As a 17-year-old, I was completely confused throughout that experience. I knew him as a celebrated physicist, a parliamentarian and an exponent of the music of Franz Liszt, and it was common to see grainy newspaper photographs of him at the piano. If you were raised in Bangalore, you always heard about Raja Ramanna (RR). But it was for the first time that I had actually heard him play, while I was sitting in the first violin section of the Bangalore School of Music (BSM) Orchestra, accompanying him to Beethoven's Piano Concerto 3 at Chowdiah Memorial Hall at the end of January 2000. We must be honest – it didn't go well, but we still made a beeline to shake his hand and claim in our circles that we had met Raja Ramanna the musician.

Later that year, in October, my father went up to him in a club they both frequented and ended up saying a little more than hello. He went straight up to RR and asked whether he'd be willing to try playing some music with an eighteen-year-old upcoming violinist. According to my father's account (we in the family always added a fistful of salt to his accounts), he simply smiled and agreed and asked me to come to his place that Saturday morning. I was not sure whether I was stunned because of my father's bewildering request or because of RR's quiet affirming response to this completely anonymous man. Over the next four years of my (literally) everyday association with him, I realised there was nothing to be bewildered about. That was RR's hallmark. Humility and warm invitation. In his autobiography *Years of Pilgrimage*, he writes about how the Maharaja of Mysore invited him, as a 12-year-old in 1937, to play the piano and enquired with him with genuine care about his music education[1]; I believe some part of RR's humility towards young people might have been inherited from such events in his own life.

What I write about in this account is about Raja Ramanna the musician in his engagement with a teenage boy, a view that few were privileged to have, and how he was my Smiling Buddha for 4 years.

Beethoven

I went to his house, very close to where we lived, that very Saturday. I felt like a long-lost uncle, not an iconic scientist, had opened the door and led me in towards his well-appointed library that housed his piano, and I hadn't the smallest inkling that he was going to be my patron for the next four years. After a second of silence and staring at him, he asked me what I'd like to play – I said why not Beethoven's "Spring" Sonata? Yes of course – was the reply. He took the sheet music from my hand, and we began to play Beethoven together.

His arthritis was difficult to watch. No wonder the concert in January that year did not go as well as expected – he was rooting for Beethoven to conquer his non-cooperating fingers at the piano. However, his difficulties turned out to my advantage as he played the piece at a slow tempo, allowing me the luxury of an unhurried accompanied practice (a boon I doubt even professional violinists have). It took that morning's meeting to realise we actually sounded "natural together" (in his own words – I was too stunned to process anything during those twenty minutes). He asked me what my phone number was. I told him (in the second sentence I spoke to him that morning) that I didn't have a phone since my family gave it up

---

[1] Interestingly, the Maharaja had requested him to play, a few weeks after this meeting, the Beethoven Piano Concerto 3 with the palace orchestra under a German conductor, the very same concerto that I mention at the beginning of this chapter. RR writes that this did not proceed well – perhaps that Concerto has had a mixed relationship with him!

a few years prior. He immediately facilitated a telephone connection to my home since he otherwise "couldn't call me up for daily practice". In a few weeks, the telephone arrived, and he actually did stick to his word and called me up literally every other day with that one word "...coming?" I was even more confused than before why this was happening.

We played almost every evening after I returned from college, for years. Beethoven's Spring Sonata, at half tempo. One hour on average, helping each other rectify difficult phrases and repeating sections we both thought worthy of playing over and over again. Over the next few weeks, he introduced me to his friends – admirals to ambassadors, disempowered rajahs, scientists of national importance, retired bureaucrats and more. Some of them visited him to just hear us practice, while sometimes the practice sessions were timed with their visits – "Anant, can you come to my place right away? [So-and-so] is here". All of them maintained pin-drop silence with rare whiskeys and cognacs in their hand, while RR and I went through movement after movement of that Sonata.

This was not just some romantic honeymoon phase of a hastily arranged musical marriage; it was a genuinely happy companionship – this incredibly decorated individual in his late seventies, I just eligible to vote. It was during this first phase of our association that I realised why he played with me, why the telephone, why the frequent dinners at club, why Beethoven and why for a long time nearly *every day*. I overheard him telling someone that this relationship allowed him to practice music on a daily basis without any pressurising commitments of a concert, or the routine pestering that many musicians brought upon him to connect them to someone in some ministry for a political favour. His family mentioned that he was actually addicted to this regularity and intensity. For me, his featherlight demeanour allowed me to bloom in Beethoven, and for an eighteen-year-old, it was a breath of fresh air to deal with a septuagenarian with no nonsense of "our good old days when a house cost fifty rupees and when people were more morally upright than you youngsters these days". He had none of that. It was purely music and little else. In December that year, he autographed and gifted me a 1939 edition of the complete Beethoven piano-violin sonatas that he owned as a student, one of the only relics of his that I still have and treasure (Fig. 1).

A few months later, in early 2001, he took a trip to Europe and returned with the sheet music of the violin concertos of Bruch,[2]

---

[2] Pronounced "brookh".

Mendelssohn and Beethoven. Long before instant PDF downloads and illegal websites offering free sheet music, laying your hand on these scoresheets was an experience by itself. He asked me to choose – I went of course with the Bruch, but it was too tempting not to try out the Mendelssohn, so I took them all home in pouring summer rain that evening. We endlessly played these pieces evening after evening. Our slow practices allowed both of us to learn these pieces at the slow pace each of us wanted, and we ended up in a harmonious joint learning exercise. Over three months, we had a repertoire ready. "Now when anyone asks us to play, we know what to offer". I cannot count the number of times we played the Beethoven Spring Sonata, the Bruch violin concerto and the Mendelssohn violin concerto – sometimes all three pieces in a solid ninety-minute recital. Of course, all these recitals had his own renditions of Liszt, with my sitting beside him turning pages. We endlessly played together at private recitals at his own home, often at his daughter Nina's home and sometimes at the homes of friends such as Admiral Stan

Fig. 1. A 1939 edition of the complete Beethoven piano-violin sonatas Dr. Raja Ramanna owned as a student.

Dawson who owned a piano. It became a regular routine for me to wait at his home, or outside of his office at the National Institute of Advanced Studies (NIAS) and then for him to take me for that evening's soiree.

Delhi

In July 2001, he asked me to travel with him to New Delhi for a series of home concerts. My mouth said yes before my mind did, and in two weeks, we travelled together. The support that his secretary Manish Chauhan and the hospitality that his staff provided for two full weeks at his bungalow at Pandara Park was unbelievable – a room for me and free access to his small but rich library at his home. We played for a full gallery of stalwarts, which expanded to people such as Mark Tully, who were well outside of the scientific and bureaucratic establishment. I very fondly recall the regular company of Dr. R. Chidambaram, Principal Scientific Advisor and a nuclear physicist of nearly equal stature as RR, who became a permanent fixture in every home recital we delivered at Pandara Park and at every other dinner at the India International Centre. Streams of visitors came and went, with me practising in the room upstairs all day and counting (but unable to name) the dozens of species of birds in the neighbourhood, besides touching the signatures in autographed books by great scientists in his library.

Staying with him for a fortnight in New Delhi allowed me to spend more quality time with him discussing music and discovering what topics never to raise with him. I quickly realised that we had much in common, for instance, in our acknowledgement of Bach as *the* supreme figure, Beethoven as a defining colossus and Mozart as something RR did not entirely enjoy though he studied his music very seriously. "Formula music!" He would exclaim disapprovingly when I tested the repertoire waters worrying whether he'd ask me to play Mozart next, a composer whose music I too greatly admired but did not entirely enjoy. He spoke at length about how young people my age were increasingly taking up courses in science and engineering only to move to jobs in finance and investment banking, which, according to him, was not only awkward for them but injurious to the scientific and technological health of the country. I picked up this a few years later during my MPhil dissertation at the Centre for Development Studies (CDS, Trivandrum) where I studied the career choices of Indian Institute of Technology Madras students at the time; I dedicated the work to him – the dissertation was eventually published as a paper in a peer-reviewed journal. Getting more confident in trying to find further common ground, I even ventured into talking about pacifism, something he did not respond to all, pretending he didn't hear what I said. But he did hear very well when something was off the mark in the house or at work in his office; I once overheard him remarking to someone else that he did not have a private car in New Delhi and was unconscionably using the official car for his musical engagements too. He was not bragging by any measure. For four years, I had not seen him even once travelling in his official car in Bangalore for our musical journeys or to the club, and he would clearly demarcate which one was for which.

We once passed Parliament House where he had just returned that afternoon from Session, and I asked him quite naively whether he had actually interacted with the then Prime Minister. He chuckled and said that he knew someone who had shaken the hand of every Prime Minister and President of India, besides Joseph Stalin and Nikita Khrushchev,[3] Saddam Hussein, Bill Clinton, scientists such as James Chadwick and many more on that list. My naivety ascended to further heights when I asked him who that was, only to receive his own hand in mine.

Music

The Indian classical music system and the Western art music system (colloquially called "Western classical music") are similar at the rudimentary level in their recognition of a basic series of notes of different pitches and the basic quadruple beat. Immediately after that, they diverge significantly. Their sound, performances, recitals, pedagogies, academia, documentation, instrument techniques and musical structures are all strikingly distinct from each other, which fuel mutual admiration and satire, constructive and ridiculous comparison, and sometimes even family feuds in households where both systems are enjoyed.

While the Indian classical music system places primacy on melody and accompanying rhythm, the Western art music system is based on harmony. Most readers of this book would be no strangers to the first system,

---

[3] He mentioned that it was during his stay in Leningrad in the Soviet Union that his daughter Nina was born; he wished to name her "Lenina", but, in his own words "...the family would have none of it, and we decided on 'Nina'".

in fact may be well versed in it, and hence I will not proceed to elaborate on that system further except to state that there are few systems of music in the world that can parallel the Indian tradition in melodic sophistication. Percussion instruments in Western tradition pale in comparison to the granular intricacy and acoustic possibilities offered by their counterpart instruments in the Indian tradition. Likewise, the length and intensity of improvisational expanse in the Indian tradition (particularly Hindustani music) which test the musical and technical prowess of performers, and their accompanists would make even the most accomplished artists in the Western tradition drop their jaw in wonder and admiration (perhaps jazz is the only genre in Western music that comes close). Form, structure, dynamics and repertoire – all unimaginably advanced and evolving in form by the contributions of legendary musicians and teachers, taking melody, rhythm and improvisation to stratospheric levels of ingenuity and complexity. Musicians such as Yehudi Menuhin in the previous century recognised this and set about seriously collaborating with Indian classical musicians to produce innovative creations that borrowed from and represented both traditions robustly. As the Indian tradition has demonstrated the pinnacle of human melodic, rhythmic and improvisational mastery, recitals and concerts of the Western tradition (often justifiably) appear to the regular Indian listener as straitjacketed and plateaued, particularly in regular mainstream recitals of Western art music which are, I agree, routinely stiff, often even sterile and lifeless. I cannot really blame the high patriots in the Indian tradition when they scoff at these aspects of the Western tradition. However, criticisms against the latter tradition such as for having "just two ragas" reflect a wilful ignorance, as those who level such accusations do not wish to understand that the major and minor scales in that tradition are not "ragas", and for that matter, the Western tradition does not base itself on a raga system *at all*, so it has not "just two" but *no* ragas in the first place. In ethnocentric intoxication, these stalwarts miss the whole point – harmony. Where the Indian tradition has barely made progress, the Western tradition has created its own universe. Harmony, orchestration and the flowering evolution of both across the last four centuries are where the Western tradition shows off its plumage and soars in creativity.

Harmony (not in the conversational sense of being "harmonious", "pleasant to the ear" or "amicable") is a fundamental feature of music where two or more notes are played in combination, that is simultaneously, to create a composite sound that acquires a tonality and character greater than the sum of its constituent notes. Often, these note combinations are referred to as "chords". Any number of notes in combination can produce harmony, though it is usually unwieldy to add more than a required number of layers – seven or eight at the very maximum – into that sandwich. Combinations of two, three, four or more notes are provided with identifiable technical names (minor third, major triad, diminished seventh and so on) and are often verbally referred to during training, performance, composition and teaching. Some combinations produce sounds that are bright, some produce effects of melancholy and despair, and with no exaggeration, I can even add to that list a spectrum of tone colours that evoke hope, nostalgia, satire, stillness, fireworks, fear, ominousness and rectitude – an endless list. And it does not stop there. The Western tradition has in its treasure chest an unimaginably diverse set of instruments – bowed strings, plucked strings, woodwinds, brass, percussion and keyboards – each and every one of these categories of instruments possessing a unique recognisable texture, character, range (low or high pitched) and acoustic possibility. The Western tradition famously follows a system of written and codified notation and a strict fixed sense of pitch. For instance, the note "middle-C" is exactly the same pitch on every instrument anywhere in the world and hence a rendition of a particular piece of music by anyone anywhere would be in exactly the same pitch. In the Indian tradition, however – and I do not pose this as a shortcoming – each person sings or plays the same piece of music at a different pitch that is optimal to one's voice and instrument's tuning.

As harmony can itself create a musical passage (such as blocks of chords like those played on a church organ) or accompany a melodic line, these note combinations have been exploited to their highest and deepest limits by the Western tradition – every few decades breaking those limits and discovering new sound worlds. I leave it to the reader to imagine then the universe of sound and harmony in the Western tradition given the infinite harmonic possibilities communicated by a treasure trove of instruments. And not just Western *art* music – rock, film music, popular music, religious music, heavy metal and even small jingles – all are based on systems of harmony. For the unfamiliar reader, may I advise listening to *Prelude and Liebestod* by Richard Wagner (a personal favourite of RR) to experience a small sample of this universe? When one enjoys listening to music from this system, one is not simply enjoying the "tune", but a thick composite sound of multiple notes at various pitches that create an experience much the same as savouring a wonderfully delicious dish which, like harmony, is about tasting a composite of various ingredients all put together in various measures and at the right moments of preparation. Those who enjoy deep listening or are familiar with performing in the Western tradition have had transcendental experiences with harmony that resonate sentiment which has no equivalent words in known human language.

RR believed that the Indian and Western traditions could speak to each other at the atomic level – an appropriately strange analogy given the protagonist here. In his short note *The Structure of music in Ragas and Western Systems* (published in 1993), he attempted to display in notation the various rhythms (*talas*) and modes (*ragas*) for especially the benefit of practitioners of the Indian system. But he, like me, however, believed (and often voiced) that a marriage between them at an advanced level was awkward. We would both always smirk when someone brought up the topic "fusion music" during a home recital – in our shared opinion, this is at best an amusing combination of both traditions and at worst a forced intercourse between two completely different oceans, procreating freakish sounds that did complete injustice to both traditions. Still, RR believed strongly that at the foundational level, several fundamental structures were alike (if not exactly the same), and that they could indeed speak to each other at that level, to the extent that at a primordial level, one could even listen to Liszt with the lens of the Indian tradition. Yet, even though he accorded serious listening to especially Carnatic music and had informed interactions with icons in that field, his heart and soul were still in the Western tradition.

And within that tradition, though RR deeply respected and played Bach and his music in the Baroque tradition (a period dated around the mid-1600s to the mid-1700s CE), he loved and preferred Romanticism (a period which took form soon after, and due to, Beethoven in the early 1800s right across the entire century). He had mentioned several times about how he enjoyed listening to great conductors (I was particularly stunned that he had heard Bruno Walter, William Furtwangler and Richard Strauss conducting) and how he considered Yehudi Menuhin as a "youth icon" during his years of study in the UK; Walter, Furtwangler, Strauss and Menuhin were for me legends of an era gone by, and RR turned out to be my direct connection with an important era in the evolution of performance in Western art music.

His hands would almost automatically rest on the piano keys, instantly ready to take off on the music of Brahms, Rachmaninoff, Schubert,

Schumann and of course the composer closest to his heart – Franz Liszt. Liszt (1811–1886, Hungarian) was himself a towering, pioneering and stellar figure in the history of piano playing and changed forever the way the instrument was played. If left to himself, RR would keep aside the piano works of even Beethoven[4] and prefer to play Liszt. He often mentioned that Liszt's music provided him an entry point into the elemental nature of music itself, facilitating a pathway, at that level, to a grand landscape view of the vast expanse of the nature of the universe itself.

It was rather amusing sometimes how, when I tuned my violin to his piano to initiate every practice, he would forget that I was tuning, and he would begin a series of sweeping harmonies and melodic lines; it took him a minute or so to return from that world of swirling Liszt-like piano passages to realise that I had finished tuning and we needed to begin our real pieces! His approach to Liszt was quite different from how even acclaimed exponents might approach and render a piece. RR would dwell sometimes on certain chords and deliberately slow-down passages that he felt needed far more attention and indulgence than what either the composer himself, or even well-known performers, thought. Without changing or omitting a single note, he took liberties with tempo even in works of early Romanticism which are rather strict about tempo and for that matter even altered tempo while accompanying me, which sometimes threw me off track, and I was met with a tiny sly smile on his face, knowing very well what had just happened. He once even indulged in some tempo stretches in piano transcriptions of pieces by Bach,[5] but of course, with due respect to the master, RR experimented only during phrase endings, in a treatment that some conductors in the past such as Leopold Stokowski also accorded to Bach in their orchestral transcriptions.

While not composed with that purpose, the phrasing structures in Liszt and Schumann did allow him to take that extra liberty in engulfing within the harmonic structures and experimenting with the dynamics of the melodic line which musical orthodoxy would scorn upon. Admittedly, I, too, had wondered whether he was playing with the elasticity of the melody and lingering on certain chords due to slowing down by virtue of age, or his finger-stiffness, or any other factor. But the ease at which he

---

[4] Despite his opinion that Beethoven's Piano Concerto 4 was the greatest work for piano ever written.

[5] Such as in his memorable rendition of Bach's organ prelude and fugue in A minor, transcribed for piano by Liszt.

tackled faster passages, and that there was no smudging of notes or chords, put any such thought to rest, and I realised he was practically conducting experiments with different tone colours because his engagement with Liszt was entirely one-to-one, without any coercion to placate the world outside. That is, when engaging with a piece of music, performers are usually pressurised – even when they become superstars – to keep up with some minimal expectations of musical orthodoxy. RR had no such obstacles, and it was just him, his piano and Liszt. I sometimes thought the musical outcome of his renditions of Liszt was akin to the trio (RR–piano–Liszt) stranded on some island, with no rigid social norms and expectations to oversee their alliance. For someone watching like me, this trio in their deepest moments in his library, with no audience or stage, it was nothing short of listener's bliss.

Home, Office and Everywhere in Between

Some of our concerts will remain in my memory very fondly. One of them was at the NIAS, for a conference in January 2003 on "Science and Spirituality". We rendered our usual repertoire of the Bruch violin concerto, with a hall packed with people who came up later to meet both of us in person. Once, and just once, I kept him waiting for me on stage at this performance at NIAS. I was waiting backstage to be called upon when I spotted none other than Jane Goodall, the primatologist, walking into the hall. I dropped everything, went up to meet her, got a lovely, autographed postcard from her, snapped a picture with her and ran backstage only to see him a little surprised. I quickly apologised to him, but in my mind, I forgave myself given my highly genuine reason for the thirty-second delay. But while I recognised Jane Goodall, my complete ignorance of some celebrated scientists in that room came to light the next day. My friends in college at St. Joseph's, who were studying science, saw a photograph of us playing that had appeared on the front page of *The New Indian Express*, were astonished at seeing a couple of those scientists in the audience and gasped even more when I casually mentioned that they had come up and offered a kind word of appreciation. Ignorance is not always bliss; I learnt (Fig. 2). NIAS was a place I frequented, to wait for him outside his room (or sometimes inside his room, gaping at all the charts and photographs), with his dog Chin-Chin running around the corridors (the poor creature met its fateful end mauled by stray dogs outside

Fig. 2.   Front page of *The New Indian Express*, 11 January 2003.

of NIAS). Ramakrishna, his trusty secretary, warmly welcomed me with his broad smile every time I waited for him at NIAS and became a senior colleague when I joined the institution in a much less interesting capacity many years later in September 2020 as a faculty of social science.

It was funny sometimes how some announcements prior to our recitals said that "Mr. Anant Kamath will be accompanying Dr. Raja Ramanna on the Beethoven violin concerto", but in some sense they were right. While by compositional design in concertos, the piano accompanies the violin, in this case, it wasn't just some piano. Sometimes modest (he once told an audience "He will be playing the music while I make some noises at the back"), it was indeed Anant Kamath who was accompanying Raja Ramanna at every recital. A public performance that was rather memorable – for not the best reasons – was at the Gayana Samaja in Bangalore, a reputed and respected institution for Carnatic classical music where it was announced once again that I would be accompanying him on the Mendelssohn violin concerto. RR insisted we play the Beethoven Spring Sonata as it would appeal to the audience's sensibilities much more than the high-Romanticism works we otherwise performed. The audience consisted of well-placed and established musicians in that genre but drifted halfway through the Beethoven and began conversing among themselves. The chatter started slowly overpowering the music and RR visibly began to get distracted. I was hoping he would not say anything, in the manner that he once abruptly stopped playing during a recital for the Bishop Cotton Boys' School Alumni meet, when he stood up and asked the audience to kindly respect the music. Even the first movement of the Mendelssohn violin concerto – unforgettable for any first-time listener – was insufficient to quell the babble. We continued playing until a few members of the audience finally realised we had concluded, provided us with a brief applause and instantly resumed talking again. On our return journey home, he told me that he genuinely loved the institution and many of its members were close friends but that he would not play there again.

Vibrant conversations were the norm at his home, where Mrs. Malathi Ramanna and daughters Nina and Nirupa never ceased in their warm welcoming smiles and informal ease, making their home a comfortable haven that supported a deeply nourishing musical experience for me (Fig. 3). How they put up with this scruffy teenager arriving every other day

Fig. 3. The usual practice in the library.

demands some appreciation. RR's library that surrounded his piano was full of extraordinary books, autographs (even one from C V Raman) photographs, sheet music, tapestries, curios and documents (particularly the series of *garudas* from across India and Southeast Asia) that kept me occupied during his depths in a Liszt piece.

Benediction

Our last public concert was at the Teen Murti Bhavan in May 2003 in New Delhi organised by the Delhi Music Society. It was also his last public appearance as a pianist. We played the Bruch Violin Concerto, which he chose over and above Mendelssohn or even Beethoven. He permitted me to play the Bach Partita in E major, an unaccompanied solo violin piece, where I faltered with ease on stage. However, the pinnacle of the evening was his rendition of the *Bénédiction de Dieu dans la Solitude* ("benediction to God in solitude") by Franz Liszt, a long, slow, moving composition that had a stunning variety of intense and passionate phrases, moments of divine silence, high drama and finally reconciliation with the universe. No composition could suit the occasion further or the man delivering the piece. It seemed like a premonition of his own final return to the universe, which transpired a little more than a year after this performance. He had at his disposal a gigantic ten-foot grand piano (which I struggled to accompany that evening given its sonority) from which he drew out all its power and majesty, with uncompromised brilliance. I had never heard him like that at any time in the last four years. We were all proven wrong – the January 2000 concert in Bangalore where we accompanied him for the Beethoven Piano Concerto 3 was no swan song. He played for dozens of recitals after that, and I was supremely privileged to be a regular part of it. Very sadly, though, not a single recording of ours exists or remains.

That evening in Delhi, he immersed himself and beckoned the entire hall to journey with him through the microcosmos of that *Benediction*, took wing in his finest hour and, at a numinous plane, became at one with his dear Franz Liszt.

He passed away very suddenly in September 2004, in Bombay, a year after I left Bangalore to study at the Madras School of Economics. My grandmother called on the phone from Bangalore and told me about it. I felt completely stuck as he was lying in state in Bombay, too expensively far for me to fly there. Despite his age, it was still shocking, given that I had practised with him just a few weeks prior on one of my regular monthly trips back home to Bangalore. I did not view his body; my only memories of him, therefore, are as my patron, on the piano practically every day, for four long years. I believed with all my heart that he was now in unity with the universe and with music.

Dr. Anant Kamath is an Assistant Professor with the School of Social Sciences at the National Institute of Advanced Studies (NIAS), Bangalore. Prior to NIAS, he spent several years at Azim Premji University at the School of Development, and at both institutions, he has undertaken and published research studies on technology and society. For several years now, he has been the principal violinist and concertmaster of the Bangalore School of Music Chamber Orchestra.

List of Events in Connection with Dr. Raja Ramanna Birth Centenary Celebration at National Institute of Advanced Studies Bengaluru

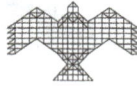

25 January 2024: Inaugural programme of the Dr. Raja Ramanna Birth Centenary Celebration

- *Inaugural Address by* Prof. C. N. R. Rao, Honorary President, JNCASR, Bengaluru.
- *Tribute Concert to Dr. Raja Ramanna – Paramanu and Paramatma* by Mr. Shadrach Solomon, Dr. Deepti Navaratna, Dr. Ananth Kamath, Mr. G. Arun Siva.

4 April 2024

- *Panel Discussion on "The Relevance of Traditional Wisdom in Modern Science" organised by* Prof. Sisir Roy.
- Panellists: Dr. Ngawang Samten, Dr. Sudha Seshayyan, Dr. Thimappa Hegde and Dr. H. R. Nagendra.

20 June 2024: NIAS Foundation Day Celebrations 2024 – Dr. Raja Ramanna Centenary Year 2024–2025

- *Foundation Day Address by* Dr. Ajit Kumar Mohanty, Secretary, Department of Atomic Energy and Chairman, Atomic Energy Commission.
- Followed by *Music Concert by Mysore Police Karnatic Band.*

12 July 2024

- Lecture Series on 50 Years of Pokhran: *The Road to Pokhran – Science and Diplomatic Perspectives.*
- Talk by Prof. V. S. Ramamurthy, Emeritus Professor, NIAS; Amb. Venkatesh Varma DB, Former Ambassador of India to the Russian Federation and Former Ambassador of India to Conference on Disarmament in Geneva.

3 August 2024

- *A Seminar on* "Emerging Technologies in Defence" organised by Dr. P. S. Goel and Dr. Sateesh Reddy.
- Inaugural talk by *Dr. V. K. Saraswat, Member, NITI AAYOG.*
- Panel Discussion by Dr. Samir V. Kamat, Chairman of DRDO, Dr. S. Somanath, Chairman of ISRO, and Dr. V. S. Ramamurthy, Former Secretary of DST.

2 September 2024

- A seminar on "Nuclear Physics Research in India" organised by Dr. D. K. Srivastava and Dr. Rudrodip Majumdar.
- Inaugural talk by Dr. Ajit Kumar Mohanty Secretary, Department of Atomic Energy and Chairman of Atomic Energy Commission.

31 January 2025: Closing Ceremony

- Dr. Raja Ramanna Memorial Lecture by Prof. Ajay Sood, Principal Scientific Advisor to the Government of India.
- Video Film: Legacies of Innovation.
- Musical Concert: Stringing East-West Resonances: The Bangalore String Ensemble.

Bibliography

By Dr. Raja Ramanna

1. "Years of Piligrimage: An Autobiography", Raja Ramanna, Viking, New Delhi (1991).
2. "The Structure of Music In Raga and Western Systems", Raja Ramanna, Bharatirya Vidya Bhavan, Bombay (1993).
3. "Sanskrit and Science", Raja Ramanna, Bharatiya Vidya Bhavan, Bombay (1984).
4. "Mukundamala of Kushekhara Alwar" rendered to English by Dr. Raja Ramanna and "Shivanandalahari of Adi Shankaracharya" rendered to English by Prof. C. V. Sundaram, Gandhi Centre of Bharariya Vidya Bhavan, Bangalore (1997).
5. "Some Studies on India's Peaceful Nuclear Explosion Experiment", R. Chidambaram and R. Ramanna, IAEA-TC-1-4/19, *Proceedings of a Technical Committee*, Vienna, 20–24 January 1975, Peaceful Nuclear Explosions IV (Vienna: IAEA, August 1975), 421–36.
6. "Moksha: A Critique, Scientific Philosophy with reference to Buddhist Thought", R. Ramanna, http://eprints.nias.res.in/433/1/L2-02_Raja_Ramanna.pdf.

About Dr. Raja Ramanna

7. "Nuclear Scientist: Dr. Raja Ramanna His Life and Work", B. V. Sreekantan, Bharatiya Vidya Bhavan, Bengaluru (2013).
8. "Dr. Raja Ramanna: A Brief Biographical Memoir", M. G. K. Menon, RRCAT Newsletter, 2006, *National Academy Science Letters*, vol. 28, Nos. 7&8, (2005); https://www.rrcat.gov.in/newsletter/NL/nl2006/issue1/pdf/N10.pdf.

9. "Raja Ramanna – A personal tribute", K. R. Rao, *Curr. Sci.*, 87 (2004) 1152.

10. "Raja Ramanna – Down the Memory Lane", B. V. Sreekantan, *Curr. Sci.*, 87 (2004) 1150.

11. "Proceedings of the Conference on 75 years of Nuclear Fission: Present Status and Future Perspective (Fission 75)", Eds. D. C. Biswas, K. Mahata, V. M. Datar, *Pramana – J. Phys.*, 85, 185 (2015). https://doi.org/10.1007/s12043-015-1073-x.

12. "Raja Ramanna: 60th Birthday Felicitation Volume", Indian Academy of Sciences, Bangalore 1985, *Pramana – J. Phys.*, 24 (1985) Issue 1–2.

13. "Raja Ramanna (1925–2004)", B. P. Radhakrishna, *J. Geol. Soc.* India, 64 (2004) 715.

14. "Raja Ramanna: Scientific mastermind behind India's nuclear tests", Haresh Pandya, *The Guardian*, October 1, 2004.

15. "Raja Ramanna: The nuclear maestro", Anil Kakodkar, *India Today*, August 30, 2021.

16. "A man of vision: Dr. Raja Ramanna, 1925 – 2004", T. S. Subramanian, *Frontline*, October 22, 2004.

17. "Ramanna and the Nuclear Programme", M. R. Srinivasan, *The Hindu*, September 28, 2004.

18. "Remembering Ramanna", P. K. Iyengar, The Hindu, September 25, 2004.

19. "Raja Ramanna", V. S. Ramamurthy, *Physics Today*, 58 (2005) 81.

20. "Fire and Fury: Transforming India's Strategic Identity", Anil Kakodkar and Suresh Gangotra, Rupa Publications India (2019).

21. "India Rising: Memoir of a Scientist", R. Chidambaram (Author), Suresh Gangotra (Contributor), Ebury Pr, India (2023).

22. "Dr. Raja Ramanna: Nuclear Physicist and Grand Pianist", Carol Lobo, Peepul Tree, August 20, 2021.

23. "There is a basic dishonesty in the system", interview of Dr. Raja Ramanna by Vandana Shukla, *The Tribune*, September 27, 1998.

24. "Dr. Raja Ramanna", C. V. Sundaram, First Annual Raja Ramanna Lecture, https://cms.nias.res.in/sites/default/filesefs/2022-01/L1-2006-CV_Sundaram.pdf.

Documentaries involving Dr. Raja Ramanna

1. "In Conversation with Dr. Raja Ramanna | Nuclear Physicist" https://www.youtube.com/watch?v=A-S3wQbTIRU.

2. "Raja Ramanna: Man W Put India on the Nuclear Power Map", https://www.youtube.com/watch?v=YPwvzKa54-A.

3. "War and Peace in the Nuclear Age; Haves and Have-Nots; Interview with Raja Ramanna, 1987"; https://openvault.wgbh.org/catalog/V_4B6BFC 12257C4824A1848B5A22285EED.
4. "Pride of India: Guru Dr. Raja Ramanna", https://www.youtube.com/ watch?v=pk5bmnickA0.

Index